CARDIFF
CAERDYDD

The Art of Flight

Fredrik Sjöberg

The Art of Flight

Translated by Peter Graves

PARTICULAR BOOKS
an imprint of
PENGUIN BOOKS

PARTICULAR BOOKS

UK | USA | Canada | Ireland | Australia
India | New Zealand | South Africa

Particular Books is part of the Penguin Random House group of companies whose addresses can be found at global.penguinrandomhouse.com.

First published 2016
001

Typeset in Bembo Book MT Std by Palimpsest Book Production Ltd, Falkirk, Stirlingshire
Printed in Great Britain by Clays Ltd, St Ives plc

A CIP catalogue record for this book is available from the British Library

ISBN: 978–1–846–14799–9

www.greenpenguin.co.uk

Contents

The Art Of Flight

Contents

To flee is to be victorious, as you know. By fleeing, you hold on to love and create a memory for yourself. You sacrifice everything for love, and all you have to sacrifice is love itself. If you remain, you will be sacrificing flight and creating a longing. Courage is needed in both cases.

Stig Claesson

I'm going to tell you about a painter. His name was Gunnar Mauritz Widforss. A watercolourist with slicked-back hair and with beautiful landscapes as his speciality. Steep mountains and big trees.

Widforss was born in Stockholm in 1879 and died at the Grand Canyon, Arizona, on 30 November 1934, at the age of fifty-five. Literally on the edge of the abyss, alone. He is forgotten. And I am his best friend.

What's more, I'm frightened out of my wits. Sometimes by almost everything, but today by the thought of failing a trust that has never been placed on me but that I snapped up of my own accord. Frightened to tell too much, or perhaps too little. Most of what I know about Gunnar I have found in letters and other hidden sources, some of which are very private. Secrets. Doubts.

Who am I to root around in his business?

Chapter 1

Nammavarejauratjah; or, Life in the Woods

Stories just begin. We rarely know where and almost never why. It doesn't matter. Nothing is certain any longer. I just want to shut my eyes, point at random and say, as a sort of experiment, that once, when I was sixteen years old, I spent a whole night singing romantic songs at the top of a pine tree. That's where it may have started.

The tree was old and growing on a mountain close to a tiny lake called Nammavarejauratjah, a couple of days walk into the trackless country of the Muddus National Park, right out in the middle of the wilderness between Jokkmokk and Gällivare. Nothing but forest and bog as far as the eye could see, for this was summer – obviously – when the nights in Lapland are lighter than on any of the midwinter days. It's very unlikely that you would be in Muddus in winter, and, if you were to be there for some reason, you would not be sitting in the treetops singing about girls with wreaths of guelder rose in their hair. Not for hours on end, anyway.

This memory came to me out of the blue one afternoon outside La Posada Hotel in Winslow, Arizona. It's significant in a number of ways, but for the moment let's concentrate on the tree itself. The pine.

If there is one thing in this world that I really understand, it is pine trees. Ordinary pines with cones and needles and a whole package of nuances: colours, scents, the murmuring in the wind in all weathers, the calls of the flocks of goldcrests, in the forest, in the autumn. Pines. I know everything about the way the sunlight behaves in a pine, and the shadows under it, irrespective of its age and where and how it is growing. Dew and mist, snow, rain – the whole lot of them, as long as they are falling in relation to a pine tree.

Please note that we are not talking science here. The purely prosaic aspects of the natural history of the pine are not things I have at my fingertips, nor are they things that interest me. I have never been a scientist. Let's talk instead about the image of the pine tree as a whole: its personality, however obscure such knowledge may sound. That's my subject, with some degree of specialization in how pine trees are depicted in early-twentieth-century academic painting. I have prowled around this theme for years, ultimately with a view to telling the story of Gottfrid Kallstenius (1861–1943), a painter disdained by the present day; and yet he was the incomparable master of pine-tree painting in Sweden. We grew up in the same town, Gottfrid and I, and the same landscape marked us both forever. During my schooldays one of his biggest paintings, signed and dated 1934, hung in the assembly hall; it depicted a solitary pine by the sea – in the sunset. That was art.

Only much later did I come to recognize that to be a fan of Kallstenius in general, and of his fiery sunsets from the years between the wars in particular, was far and away the most stupid thing you could confess to in terms of art. Anyone in the least afraid of being cold-shouldered was best advised not to mention his name except, perhaps, as an example of the mildewed fustiness that the fresh breeze of Modernism was supposed to have blown away once and for all. So I put up with the pseudo-Matisse paintings that everyone else had deemed acceptable.

Poor old Gottfrid – he lived too long, I think. If he'd drunk himself to death at the turn of the century or spent the second half of his life in a lunatic asylum, liberating his soul by producing childish scribbles on poor-quality paper, in the fullness of time he would have been lauded as one of the truly great. But no. When he was at his peak – on a par with Anders Zorn and Bruno Liljefors – and had been elected to the Academy of Art, he bought a house by the sea in Källvik and began to paint pine trees in evening light. He carried on doing so for forty years. Most of these are bad, but a few are good. And the occasional canvas is magical.

So I decided to be bold and to set off on an expedition. My intention was to start travelling again after all my years on the island I rarely leave: Helsinki, Munich, Budapest, Boston, Indianapolis, Buenos Aires and all the other places in the world that have great museums that hold paintings by Gottfrid Kallstenius down in their darkest underground storerooms. The plan was to make a study of the morphology of failure.

Bloody hell, I thought.

Not much of a thought, I admit, but nevertheless it's the thought that counts. Gottfrid and me – and a thousand pine trees.

My comeuppance was all the greater. I gave up, crestfallen and disappointed. Sweden is a small country and it didn't take me long to discover that a claim on the territory had already been staked by an art expert with a good deal more knowledge than I had and with established connections to the artist's relations, along with everything that implies for access to journals, letters and the other treasures a researcher likes to work through in solitude. This was a harsh blow, and I won't deny that I hurled the occasional unjust comment over my shoulder – something about the sterile old sea eagles of the 1980s, hopelessly damaged by DDT but still occupying all the best territories in the archipelago and consequently blocking the breeding chances of the younger birds that had grown up with less poison in their bodies.

So the best nesting trees were out of the question, and the young birds had to nest as best they could in the straggly seed-pines that grew on the clear-felled, wind-swept blocks. At the end of May, which was when we used to ring the chicks, the nest would frequently lack any sign of life and be little more than a heap of brushwood on the ground. Eagles are no good at building, not at the start, and they do best in inherited nests.

This happens sometimes. You find yourself a painter, you take control of the territory, secure the unique documents and bide your time. Though, to be honest, it has to be said that in the case of Kallstenius the fault was my own. I didn't dare. The field had been clear and free for several years and no one else was bothered, while I just talked around it and thought half-thoughts. And by the time I finally made up my mind it was too late. Lucky for Gottfrid, perhaps. Anyway, I'm only telling you this to explain why I was so quick off the mark in the case of Widforss and why I made a hasty decision I might have been more careful about had I given it greater consideration.

What happened was this.

On Saturday, 29 January 2005, the whole family was on the mainland. We had booked a window table in the restaurant of Moderna Museet in Stockholm. What had brought us there was Kazimir Malevich and his famous painting *Suprematist Composition – Black with a White Rectangle*, painted in 1915. The museum had owned it for the best part of a year, but the restoration of the canvas, which had been rolled up for a long time and was in poor condition, had proved to be time-consuming, so it was not until January that it went on display. A mystery. An infected wound. Not the painting itself, which is Russian avant-garde and no more than that, but its story, the legends, the half-muted songs and the quarrels among art historians. The little I knew of it sounded like the plot of a novel. Was it really stolen? Secretly I hoped so.

We'd arranged to meet our children at the museum.

We were to meet there late in the afternoon, look at the painting and eat afterwards. Johanna and I had time to spare and decided to drop into Bukowski's Auction House down by Berzelii Park on our way to the museum on Skeppsholmen in order to view and possibly bid on an early painting by Mollie Faustman (1883–1966), one of the first Swedes to study with Matisse in Paris as well as being – much later in life – my wife's godmother. She went to Paris early on – 1909 – after a bizarre upbringing with an unhappy father who was the product of an extramarital romance between Lars Johan Hierta and Wendela Hebbe. It's a story that surpasses a good many fictional tragedies.

Hierta, best known nowadays as the founder of the newspaper *Aftonbladet*, was a splendid fellow with a wife, children and house in Gamla Stan in Stockholm. When Hebbe, the first woman journalist in Sweden and Hierta's mistress, became pregnant she was despatched to Berlin on some obscure mission, and there she gave birth to a child who was immediately given up for adoption and taken by a family called Faustman. That's the way they did it in those days.

Hushed things up. As ill-luck would have it, however, Hebbe, who quite astonishingly also happened to be a lodger in the Hierta household, regretted the adoption a couple of years later and wanted her son back. That's understandable, of course. The strange thing is what happened next: Hierta adopted his own son, the Faustman boy, but did so without telling him either who his real father was or that the kind lady upstairs was his mother. Given all this, it is hardly surprising that the man who was to become Mollie's father was a troubled soul.

There was something wild about the large painting – *Recumbent Model with a Blue Book* – but we liked it anyway, or perhaps it was Mollie Faustman we liked, I'm not sure. Something, anyway. And since we were there for a viewing, we took the opportunity to have a leisurely browse through everything else that was up for auction. I spent some time standing in front of a gigantic sunset or, more accurately, moonrise by Kallstenius, dated 1930, which – with a gentle glow of *Schadenfreude* in my heart – I found to be utterly worthless. The breach between Kallstenius and me was of quite recent date, so everything was still a touch fraught.

That was the kind of mood I was in when I suddenly caught sight of a pine tree. An ancient pine tree by the sea, crooked and stunted and painted by an artist I didn't know: in the midday sun of high summer, straight on, no low-angle lighting or symbolism. And what I saw I saw immediately, in a fraction of a second. What a find! The catalogue stated 'Gunnar Widforss, Sweden/USA, 1879–1934, *Pine Tree at Roskär*. Signed Widforss and dated 1917. Watercolour, 45 x 63 cm'. Estimate 3,000 kronor. 3,000!

Experiences of art can actually afflict you in rather the same way as falling in love, in the tiresome and usually unwelcome sense that a lust for possession is lurking in the shadows. Sad, no doubt, but if you hang around auction rooms you have no one to blame but yourself. Not that it was something I gave any thought to on this occasion. In fact, I didn't actually think at all, hardly gave myself time to look at the painting before chasing around for a while trying to find Johanna, who had become transfixed by, to my eyes, a meaningless

Modernist. She seems to dislike works of art that depict fir trees almost as a matter of principle, but there are some things you really want to share, so I took the risk.

'Are you going to buy that thing?' was the first thing she said.

. . .

Those of us camping by Nammavarejauratjah in Muddus in those light summer nights weren't exactly a troop of baboons, but we weren't far off it: a gang of teenage birdwatchers spending a couple of weeks hiking through the forests, across endless bogs and – perhaps above all – discovering one another. A great deal is said and sung in trackless terrain. And I've no doubt that it was only from within the warmth of the group that the landscape became really living and visible. Without that security it would never have occurred to me to go off on my own and climb up a pine tree one night. I shall never forget it. The wilderness. The scents. The songs! Those of the redwing and the rustic bunting, too. Perhaps that's where it started.

Chapter 2

Dear, Dear Mamma

Showing Moderna Museet's new Malevich to half-interested teenagers is a challenge up there with attempting to explain what Jackson Pollock was thinking about when he was thinking, which he may well not have been doing when he executed the panel that hangs in the adjoining room. It's not easy but, with a little patience, it can be done. The young people seem to have an instinctive feel for anything and everything that provokes and oversteps the boundaries. They were familiar with Robert Rauschenberg's goat stuck in a car tyre before they'd started school, and since then they have been sufficiently broad-minded to appreciate the humorous inventions created in high flights of intoxication.

The retrospective exhibition of paranoid installations by Ann-Sofi Sidén on show one floor down was more difficult to digest than Salvador Dalí, Jean Tinguely and the other old troupers, but it served to round off my afternoon, which had started on the unfashionable outskirts of art but slowly and surely approached the centre. No one can take the pulse of the present day in the way Sidén can, and it seems to me that there is nothing else around at the moment that corresponds to the academic painting of the turn of the last century in the way her work does. The style is different, but the public role is of the same kind. Her 1995 installation, which consists of a doorway blocked up to the last millimetre with textbooks on psychiatry, is eminently worth seeing and feels as if it's in a direct line of descent from Gottfrid Kallstenius's *Moonlit Night on Gotland* (1900), a work of similar brilliance and no less saturated with symbolism even if it is rather achingly National Romantic. The latter was snapped up by the National Museum almost before the varnish had time to dry.

For Gunnar Widforss the road was a longer and dustier one, as I was soon to discover. But, then, he was not bobbing along like a champagne cork in the mainstream of the art world of his day. There were even some people who said that he wasn't an artist at all.

For my part, however, after just a couple of days I was hopelessly trapped in a maze of loyalties, confused by gold fever and the elation of the treasure hunt – as I always am when I first pick up a promising scent that might lead heaven knows where. The date of the auction at Bukowski's was not until 8 February, which meant that I had ten days, a veritable ocean of time, to devote to preparatory studies and get myself worked up. I hungrily devoured what little had been written about Widforss in what sources were available to me. It was enough – more than enough.

The result was that by the time the auctioneer was standing there clearing his throat I knew that Gunnar Mauritz Widforss had not studied at the Academy but at the Technical Institute, the forerunner to the present-day College of Arts, Crafts and Design, where, in 1900 at the age of twenty, he qualified as a journeyman decorative painter – not a real artist, then – after which he seems to have spent the rest of his life as a restless spirit constantly on the move. He travelled in Russia, Switzerland, Austria, France, Italy, Germany, the USA, Tunisia, England, Holland, Hungary, Denmark, and then again the USA, where, at the beginning of the 1920s, he finally stayed, eventually dying there – to be absolutely precise, in his car. In his adult life the only continuous period of any length he spent in Sweden was during the First World War. That explained my pine tree, or what was soon going to be my pine tree.

As the ordinary clientele flocked into the auction room I could bolster my confidence with the knowledge that Gunnar, who remained single and childless, had achieved success with a couple of watercolours at the 1912 Salon in Paris, that he had obviously had some degree of contact with King Gustav V and that he was eventually very successful in the United States as a wilderness painter. By that stage I had learned off by heart the final sentence of his brief entry in the *Dictionary of Swedish Artists*: 'In honour of "the painter of the national parks", as he was known, in the autumn of 1938 the United States Board of Geographical Names decided that a 7,800-foot mountaintop in the Grand Canyon National Park in Arizona should be given the name "Widforss Point".' More than that I simply did not need to know. My mind was made up and I was on my way.

The show could commence.

I've often wondered how long auction houses keep the video recordings taken by the camera at the back of the room in case there are any disputes and misunderstandings about the bids. I could imagine the films being archived, cut and joined, the best bits anyway, and shown at the annual office party. In which case we can be sure that Lot No. 526 on 8 February 2005 is already a classic that has made the sorely tested staff of the place laugh themselves silly. The spectacle might also be of interest to a historian of off-the-peg clothing, since for some reason the majority of speculators in art wear such incomprehensibly ill-fitting clothes that an outside observer cannot avoid the suspicion that someone at the heart of their circle picked up a job lot of suits from East Germany. They are made of a kind of synthetic industrial waste product, possibly wood fibre, that goes so stiff in its refined form that the suit can be leaned against the wall rather than hung on a hanger in the cheap hotel rooms in which I imagine men of this calibre spend the night. Perhaps, I thought nervously, it's just that art dealers are unusually thrifty.

Anyway, the bidding opened at 7,000 kronor, well over twice the estimate, and a few seconds later it had reached 20,000. Someone raised his hand for 21,000, at which a man over on the left of the room called out 30,000 in a voice as loud and resolute as that of a harbourmaster up north in the Gulf of Bothnia. There was a moment of hesitation around the telephone table and then came 33,000. I was lost. Not the chap from up north, though, who without a moment's hesitation bawled 40,000. Even the auctioneer was put off his stride. The girls on the telephones waited – but no. The hammer came down. The man got up and left. It was all over in less than a minute and I hadn't even put in a bid.

My life was ruined. I can, of course, see the pathetic side of it all, you can be sure of that, but I simply could not control my despair. It was as if I had been knocked to the ground. Crushed. After all, the watercolour was already mine! And the story! I remained sitting there for a while – a few minutes that, I imagine, must look particularly good on that film. Then I left, too.

The man was standing at the counter where buyers collect their purchases and was busy wrapping his painting in corrugated cardboard. He did not resemble anyone else who was there. Tall and well built, a good fifty I should think, hair with streaks of grey, jeans, leather jacket, boots, studs in his belt. Studs in his belt! I cast a last envious look at the pine tree before it disappeared into the wrapping – forever. Gone. At the last moment, however, I summoned up the courage to introduce myself to the man, congratulate him on his acquisition and tell him I was working on a biography of Gunnar Widforss. I admit that was not true, but what else could I say? I was certainly considering the idea and so I enquired about the possibility of photographing his painting for eventual reproduction in the book.

'Would that be possible, do you think?'

'It's not for me to decide,' he said. 'It's going to the USA. I've already sold it. For $30,000. I'm flying to Phoenix next week.'

He disappeared out to Arsenalgatan.

It was only then that I finally admitted to myself something I already knew but that I had suppressed as far as I could: nowadays Americans are prepared to pay sensational sums for Gunnar's paintings. A good watercolour might fetch hundreds of thousands of kronor; the record is over a million, the sort of sum that no other Swedish watercolourists apart from Zorn and Larsson have ever come near. What had I actually been hoping for? Only a twitcher who has arrived too late to see the bird can wallow in self-pity in the way I did on that occasion.

. . .

We were in Beijershamn Nature Reserve when the alert came. Immediately after sunset. A solitary sandpiper, north of Gårdby, on the eastern side of the island. The first ever sighting in Sweden. Could we get there in time? It was fifteen kilometres away, across Stora Alvaret, Öland's great limestone plateau.

It was the end of May, and we were on Öland as usual. This was 1987, before mobile phones, but even at that time there was already an automatic telephone answering service you could ring and get the latest news about the rarest accidentals. Where and when. We ran to the car and, after a crazy journey, managed to arrive in one piece as the twilight was deepening. The main road looked like a motorway after a multiple pile-up. Dozens of cars parked haphazardly, doors and tailgates half open. The bird was reckoned to be in a bog a couple of hundred metres into the wood, or so they said. It would soon be dark. Clutching our telescopes, we ran as if our lives depended on it.

And made it. We saw it. We saw it just before the bird disappeared into the May night. A solitary sandpiper. The first and, even now, the only one. Fifty or so birdwatchers sort of glowed in the darkness.

The point was that some people got there too late. Some of the best, some of the real ornithologists, those who had been out on the meadows along the shore that evening, or standing in tower hides spying out to sea as long as there was enough light for their telescopes, missed it, whereas people like us were hanging around telephone kiosks with our engines running. I can remember their misery. Sorrow mixed with shame, because you can't really grieve seriously over a bird you never saw.

Hope lived on through the night but, as expected, was gone by dawn. In spite of which there were grown men who stayed all the next morning, all day, all evening on the edge of that birdless bog north of Gårdby. In dogged silence.

. . .

No, I wouldn't want to call it sorrow, but there was undoubtedly an empty space into which the groundwater of shame could seep. But it didn't take more than half an hour – time for an aimless stroll around the Nybroplan area – before my self-esteem began to return and what dominated my emotions was fear of losing the biography as well. I half ran to the Royal Library in Humlegården, where, as always, the staff in the manuscript section were extremely accommodating.

'Widforss?'

'Widforss.'

'Same spelling as Mauritz Widforss AB, the hunting equipment shop?'

'That's the one. Mauritz was Gunnar's dad.'

The friendly librarian disappeared into the inner archives and was away for a while. I scribbled 'the Firm' on a slip of paper and then added 'the Family'. In the phone book, which I had already checked, there were a number of people with the name, and it wasn't unreasonable to assume that they might be in possession of some information about Gunnar's life and career. And then there was whatever might exist on the other side of the Atlantic. Books from second-hand bookshops in California and New Mexico were already on their way. The only question was, was there anyone else on Gunnar's track ahead of me?

The fellow who had disappeared into the archive was taking his time. There was nothing I could do but sit there and hope for the same sort of luck I'd had in the same reading room on a past occasion, when I'd been skimming through a bundle of letters and postcards from an unknown man in Rio to his almost as unknown father in Stockholm and as a result had managed to unravel the story behind a mysterious dedication on a painting of an unknown landscape by J. A. G. Acke (1859–1924). In fact, luck is perhaps the wrong word, because I'd actually been searching in a pretty single-minded way for months. Good luck presumably operates in the same way as bad luck, at least in the sense that we can lay ourselves open to it. Not infrequently we set up bad luck days for ourselves, I thought, after the librarian had at last reappeared and announced that he had gone through every database of manuscripts held by the Royal Library without finding any trace of the name 'Widforss' – no letters, nothing.

'Perhaps the hunting shop has a company archive?' he suggested.

It did not take me long to ascertain that it didn't. The family shop on Fredsgatan still bore the name Mauritz Widforss, but other owners had come and gone over the years, and no one had heard of a painter by the name of Gunnar. I bought a tin of air-rifle pellets and slunk away.

That left the family. I don't really like the business of snooping around among the relations of the more or less forgotten figures who catch my interest for one reason or another. I feel I'm intruding and poking my nose into things that don't concern me. What's more, I feel compelled to explain what I'm up to and why, which is something I'm not always certain about myself when I first start following a trail I think might be worth following. The most difficult thing of all is contacting the relatives of deceased artists, since they have frequently had to suffer sporadic and not always happy approaches by garrulous businessmen in wood-fibre suits.

As I've said, however, the family was my last

remaining chance. The following morning I reluctantly began ringing around and to my relief quite quickly made contact with an approachable elderly lady in Simrishamn, the daughter of one of Gunnar's brothers, who was not only an artist herself but also an acquaintance of my mother-in-law. Sweden is a small country, no doubt of that. A couple of days later I went there and within a week it was clear that no one else had got there first. Gunnar was mine. I had found myself a painter.

. . .

The family turned out to be a large one. Gunnar's mother bore thirteen children. Nine of them survived to adulthood and three of them had children of their own, who, in their turn, had gone forth and multiplied, abundantly. Those in my own generation – the grandchildren – were numerous, but they had no direct memories of the painter. There were watercolours, of course, and anecdotes and the odd letter.

No archive, though. But an architect in Vällingby, the son of one of Gunnar's sisters, had stacks of sketches, old exhibition catalogues, newspaper cuttings and photographs. It was thanks to this kind man, incidentally well over eighty, that I decided to climb a mountain. Before I met him he claimed to be slightly senile and thus unable to remember exactly what he had and where it was. That, of course, was not the case. Everything seemed to me to be in exemplary order. But he stuck to his guns. A couple of months later there was a laconic message on my answering machine.

'What did I tell you? I'm senile. I've found the letters.'

He had needed something from his wardrobe and, as we all know, architects often have big wardrobes. He was looking for something, it's not clear what, but as often happens he found something else. There was a box, rescued from the top of a skip at some point when the estate was being divided up, and it contained 360 letters and closely written postcards to 'Dear, Dear Mamma', dated between July 1901, when Gunnar began his travels as a journeyman in St Petersburg, and 26

November 1934, just four days before his death in the United States. I booked my flight the same day. Las Vegas, Nevada.

Chapter 3

The Art of Not Travelling

Most of us tend to know a little bit about everything, but seldom do we know a great deal about something, and only very exceptionally do we know more than everyone else. We usually only come top when it involves something pretty limited and insignificant, and, indeed, that is actually surprisingly easy to achieve compared to the difficulty of keeping up with all those other things that we know a little bit about or even quite a lot about. In the end we just sit there with our bits and pieces of half-knowledge and wonder what the point of it all is. It is not dissimilar to what it was like when the time came to clear out our boyhood bedroom and start life: we ended up ruminating about what could and should be done with all the beer cans we had collected during our teenage years. Bin them? Store them in the family garage perhaps? It seemed unlikely that they would ever come in useful.

The empty cans of the mind may take up less room, but they are just as troublesome and, in their own way, just as bulky. Tossing them overboard and fleeing from the past is an option that never fails to be attractive, but, in spite of that, I continue to drag them around with me, perhaps because I know what I'm capable of and recognize my limitations: if I don't hang on to what I have, there won't be anything left. It always seems to me too late to start on something new and to concentrate like a tunnel-visioned researcher on one thing in particular and follow it in a straight line. As if it were just a matter of completing a cheap jigsaw puzzle.

The role that Gunnar Widforss was to play was by no means insignificant. When authors, artists, photographers and scientists were creating the notion of the wilderness in the United States and clearing a space for it in the hearts of Americans, Widforss was in the vanguard. The painter of the national parks. That was what they called him. An utterly superb tool in the hands of those who wanted to preserve nature by appealing to emotion rather than to knowledge. And they succeeded amazingly well. Never has the politics of nature been as successful as it was during that period, the first decades of the twentieth century. In Sweden, too. What was being prescribed was an idea – to all intents and purposes untested – that virgin reserves should be placed here and there throughout the country, like Sundays in a landscape of weekdays. And, above all, what was to be preserved for posterity in the Grand Canyon, Yosemite, Muddus and Norra Kvill was beauty.

They went to work with scout-like enthusiasm and virtually religious ardour, marking out reserves both large and small as if they were cathedrals in the mountains and forests. A generation later these reserves were to be the indisputable fixed points of my own journeyman days. Every reserve a victory, every national park a shrine. That was my simple childlike faith. Until I lost it.

The doubts crept in bit by bit, I suppose, but it was on Madagascar that they finally threw me. I have never really recovered. It's a long story and I won't recount the political twists and turns here, because they did little but create animosity and turn me into an outsider, a position I had never sought.

. . .

It was on the road to the mica mines. The unavoidable preliminary was a whole day in a good car travelling through dry flat landscape on the Tropic of Capricorn in southern Madagascar, where the soil is red and the roads lined with baobab trees that look like massive stoneware gin bottles.

As usual, it all began by pure chance, in this case while sitting with a close friend under an awning outside a bookshop on the Avenue de l'Indépendance in Antananarivo. It was a very beautiful day and fresh, too, since the city is situated high in the mountains. I had just converted a very small part of my first writer's grant into a beer bearing the Three Horses label, and with that as a prompt for my memory was attempting to spell my way through one of the French-language papers of the region. It was slow going.

Then my eye chanced to fall on an article reporting that President Ratsiraka, a Stalinist who had reformed just a year earlier, had taken it into his head to decorate a Monsieur Åkesson, a man prominent in the country. He had appointed him a member of the Legion of Honour. A businessman and evidently the richest man in the whole of Madagascar. All of which would, of course, have been of no more than limited interest to me had it not been for Monsieur Åkesson's first name, which turned out to be Bertil. So a fellow-countryman.

Bertil Åkesson lived in the diplomatic quarter and gave us an open-hearted welcome and immediately offered his services. Did we perhaps want to see his estates around Fort Dauphin? Private planes, hotels, quality cars with chauffeurs? All available, just say the word. Unfortunately he didn't have time to show us his empire personally, because he was about to shoot off elsewhere on business. Besides, he had bad knees, as we heard from a French woman barely half his age who, for reasons unknown, was part of the household. He kept his wife, as we discovered later, in an apartment in Monaco.

Well, my good friend and I certainly appreciated the great man's generous offers and we were soon in Fort Dauphin, a thousand kilometres down the coast, quartered in a deserted hotel complex, at the gates of which a cheerful *garçon* would appear in a brand-new Land Rover every morning ready to drive us wherever we wanted to go. And if we didn't know where, he would drive us of his own accord to the places the business acquaintances of the Legionary of Honour usually wanted to see, mainly endless sisal plantations, which were clearly the core of his activity. That and mineral mines. Oh, and the merchant fleet. Never has my nationality – which I've done nothing to deserve – profited me more. The bosses of the various daughter companies along the way were as helpful as if we were making a state visit. We posed happily under hand-painted signs that announced that everything, absolutely everything, visible from that location was owned by

Monsieur Åkesson. And I can assure you that the view was extensive.

The thing about sisal plantations is that you quickly get tired of them, so we asked to be shown something different. That's how we ended up at the mica mines, said to be the biggest in the world, in the mountains a day's journey inland. And on the way there we passed a beautiful primaeval forest reserve, full of lemurs and rare ground rollers, which our host had bought and preserved with ecotourism in mind; even then it seemed a business with a promising future. Possibly he had more idealistic reasons in mind as well, just as had been the case in the United States a century before, and at home in Sweden soon thereafter.

The thought that occurred to me in the car that day was that in the old days missionaries used to build churches, whereas nowadays we set up forest reserves in response to a different kind of barbarism, but the reverence being exported is of the same sort. Just one more step on the road. But towards what?

The glittering mica mines lived up to the promise of their name, so my ruminations, as always, hardly had time to do more than touch the topic before they were replaced by something more entertaining and curious. The deepest shaft was narrow and black, and we felt our way along the passageways with the help of butane flames. It was damp and dark, and we were told that there were snakes that had fallen the 150 metres down the lift shaft and were still there. The biggest mica crystals are cut into thin slices the size of lunch trays and exported to the Japanese electronics industry. The rest is ground into very fine powder and shipped to every country that demands that beauty needs to be further enhanced.

'Eyeshadow, you know,' said the mine foreman, smiling in the darkness and gently stroking his eyelid with his middle finger.

'Think about it. It sort of glitters.'

. . .

There were those who said that I sought conflict in order to flee from myself, and there may be something in that, but my growing doubts about the very idea of nature reserves led mainly to contradictions within myself, for there is no one who loves these Sundays more than I do: islands, everything exactly the right size and with an outer limit. But passion is not dependent on thought, it just is.

While standing in the middle of this morass, I had finally discovered Gunnar Widforss – another fellow-countryman, also generous – whose very arrival on the scene was like a friend's personal invitation to the national park of all national parks. A piece of the puzzle, several pieces, perhaps. A journey, adventurous from the very start.

. . .

Even at a distance, the boy in charge of the exit barrier at the pick-up point for rental vehicles in Las Vegas had the look of a holiday stand-in. Spotty and cheerful, sort of tousled. He stood there with earpieces plugged in his ears, singing to himself behind the tinted glass of his booth as we glided up in the silver-grey Ford Focus we had booked on Sveavägen in Stockholm before leaving.

Johanna lowered the side-window, rested her elbow

on the ledge, pushed her sunglasses down to the end of her nose, bent her head forward a little and fixed him with a stare at the same time as revving the engine with her right foot. Just a quick burst, a marker so to speak. Too many films, I thought, but said nothing. She's the one who drives, after all. I do have a driving licence myself these days, but I got it a bit too late, in my forties, so I prefer not to use it.

The young man came to life, quickly disposed of his earpieces without any abatement of his convulsive jerks, which reminded me rather of a sweater on a clothes-line in a gale. All of a sudden he was full of bounce, saluted us, gave our documentation no more than a cursory glance and generally behaved in a ridiculous manner. It was hot. Not yet insufferably so, but hot, and the light over the city was harsh and dazzling.

Anyway, everything was found to be in order. He raised the barrier, Johanna pushed up her sunglasses and took a moment to find the right position for the automatic gear-shift. I focused on the map. Exactly what the boy in the booth was focusing on never became clear, but one way or another he must have come across one lever too many, because just as Johanna found the shift, waved to him and put her foot down on the accelerator, a rail with four-inch iron spikes shot up out of the road surface immediately under the barrier boom as if this were a covert military base rather than a civilian facility and we were enemies caught at the very last minute. That's what it felt like, anyway, once the sound of the exploding tyres died away and everything fell silent.

And there we sat.

Approximately speaking, we had covered 350 yards of our American journey, and we already had flats in all four wheels. We looked at one another in astonishment and, as tends to happen in middle-aged marriages, the reflex was to prepare to blame the other partner for what had happened. I can remember that I'd had time to come up with the semi-automatic response that this certainly could not be put down to faulty navigation on the part of the map-reader. But there was no need for me to say anything, because at that moment our friend the holiday stand-in came bounding over like a desert rat, full of apologies and almost palpable assurances that we definitely were not the ones who had done anything wrong.

'An accident! How tiresome!'

'Is everything okay?' he asked anxiously, virtually throwing himself in through the still-open window.

My pulse was pounding away so I couldn't collect my thoughts and come up with a suitable answer. To be honest, I can't remember. It sounds stupid but the only thing I noticed was that he looked considerably smaller now, as if he'd shrunk. First, however, I quickly worked out that he must have been standing on a stool in his booth and, second, he was probably by nature the sort of person who ought not to stand on stools under any circumstances. It is a fact, both well known and frequently testified to, that the experience of shock can spawn great thoughts and spur men on to bold decisions. This was not one of those occasions. Quite the reverse: I suddenly felt utterly exhausted.

Now other people arrived on the scene, clearly staff of a more permanent kind. Real men in check shirts with more excuses and more questions. How did it happen? Are you hurt? How do you feel? All those who came hurrying over deported themselves very earnestly and sympathetically, operationally business-like, strangely enough with one exception, an older fellow who looked a little redundant or as if he had not been allowed to do anything but clean windshields day after day. He just smiled – yes, really! Kicked the tyre a bit, laughed to himself and uttered words that no one would willingly quote, not indoors anyway. But it was Johanna who really took the prize.

She had recovered surprisingly quickly and was now standing in the blazing hot sunshine in the middle of the street with a circle of check shirts around her engaged in what was becoming a high-spirited conversation about how best to solve the problem. There was talk of changing the wheels, that much I heard from where I was standing a short distance away and getting all het up. Most of the morning had already gone and we had a fair old schedule ahead of us. The idea was that we would drive through the Zion National Park and pass through Springdale, the village in the mountains where Gunnar had lived in isolation in 1923, on our way. Part of the plan was to stop there and look around for a while. We were hardly going to have time for that if we were forced to hang around for these damn wheel changes.

'Tell them to provide us with a new car,' I said in a sour voice.

'Can't you keep quiet for a minute?' she said. 'I'm dealing with this.'

I gave up, let go of everything, sat down on the sidewalk in the shade, my back to a dustbin, and took out a notebook I have carried with me for many years to jot down odd notes with a view to using them in a book I shall probably never get round to writing but that might possibly be called 'The Art of Not Travelling'. In the age in which we live, saturated with experience as it is, travelling has in many ways become an overrated pleasure. That is my simple thesis, anyway, the little grain around which I am collecting evidence and reflections as patiently as a freshwater mussel, which no one is ever going to fish up (never, ever) and even less open, for what it contains is nothing but my own fears, which I have tried to encapsulate and call the truth. And now the moment had come for me to note down the observation that a good many journeys implode on day one, before lunch.

Having got that far, and just as I was about to add a reflection on the physical and emotional similarities between flats all-round and running your own boat aground, I heard Johanna in the background pronounce the comment of the day – a remark, as I realized on the highway later, that was probably the salvation of the whole trip.

What happened was that another person had joined the cluster around our wrecked car. He was obviously the boss of the business and now, after first having informed himself of the facts, he turned to Johanna and said something I couldn't hear but that made her stuff her hands in her trouser pockets and look thoughtfully first at the car, then at him and then at the car again before saying, with a little shrug of her shoulders:

'Shit happens.'

At which the man's eyes began to sparkle, and a smile of pure accord and genuine joy spread across his whole face. That was the moment, perhaps, when he realized for the first time that we were just Swedes, that we had neither the time nor the desire to claim compensation and sue him to the portals of Hades. That all we wanted was a new car and to get on our way.

'Where are you heading?' he asked.

'The Grand Canyon,' she said.

'Come with me,' he said.

They disappeared into the parking lot. An absolutely enormous tow truck was approaching down the street from the east.

All of this explains why, for several unforgettable weeks in the month of June, we drove far and wide around the roads of Nevada, Utah, Arizona and Colorado in a lime-green Ford Mustang V-6 Deluxe – the latest model.

I heard the sound of the engine from a long way off as she returned a little while later. One hand on the wheel, elbow sticking out. The film was rolling again. It wasn't a case of her having been offered the Mustang; she had been offered a different car, more like the one we had paid for. But this beast had happened to be in the same corner of the parish, parked right alongside the proffered replacement vehicle, and in her inimitable way she had exclaimed that she could think of nothing she would love more than driving a car of that colour. I can just picture the rental agent. He had lost from the start. Not a hope.

She slowed down, stopped, banged the roof twice with her left hand and yelled: 'Are you coming or not?'

I loaded the cases and was struck by the thought, as so often before, that the problem is not perhaps the travelling per se but something else, though I can't put my finger on what.

Chapter 4

Too Bright and Too Much

The extent to which artistic tendencies are inherited – genetically, I mean – is open to discussion, but whatever we feel about it the case of Gunnar Widforss tempts us to speculate. If we trace the family back on his mother's side, there is a whole series of gifted artists and craftworkers. Likewise in the generations that followed him.

The story begins with Aaron Isaac, a German stonecutter and jeweller who settled in Stockholm in 1774. Gustaf III personally awarded him the privilege of carrying on his craft without first having to give up his religion, a condition that had previously been absolute for Jewish immigrants. Somewhat later a young apprentice engraver from Prague, Salm Salmson (1766–1820), joined the business in Stockholm, married into Aaron Isaac's family and produced twelve children, the majority of whom became engravers, sculptors and painters. But it was in the following generation that the brightest stars appeared: one of them was the artist Hugo Salmson (1843–94) and another, and by no means the least, was his cousin Lea Ahlborn (1826–97), the medal engraver who in 1855 became the first Swedish woman in state employment.

Lea Ahlborn, who was the sister of Gunnar's maternal grandmother, had an amazing career. For more than forty years she worked as engraver to the Royal Swedish Mint, and during that time she created all the coins minted in Sweden and Finland as well as hundreds of medals. The Academy of Art in St Petersburg and then, later, its Stockholm equivalent made her a member, as did a number of other academies overseas, such as Boston, Philadelphia and New York. An absolute firecracker, and her elder sister

Carolina Wiedenhayn seems to have been made of the same stuff.

Carolina Wiedenhayn (1822–1902), Gunnar's grandmother, was also an outstanding engraver, a specialist in xylography (carving woodcuts) as well as being a teacher at the Stockholm Technical Institute. Anyone looking at this family chronicle can't get away from the impression of matriarchy. The menfolk are kept under wraps in the background, so to speak: if they aren't mad, they are customs officers or nouveau riche merchants sired by small farmers in Ockelbo and Döderhult. They are to a great extent short-lived figures who seldom make much of a mark in the annals.

Gunnar's mother and father are no exception to this rule. Admittedly Mauritz Widforss (1836–1905) does at least survive in the name of a shop that sells firearms and leather breeks, but during his time on earth he seems to have cut a pretty inconsequential figure beside his wife, Blenda. It does have to be said, however, that he was much given to reproduction, with the result that Blenda gave birth to thirteen children before getting started on her own, admittedly modest, career as a painter. She too had been to the Technical Institute in the past and, according to family tradition, she returned to her artistic studies once the children were standing on their own feet.

Blenda Widforss – Dear, Dear Mamma – was born in 1851 and did not pass away until 1935. Which meant that she not only outlived her husband by three decades but also seven of her children, including her dearest boy, Gunnar. And that is probably how we should see him: a Sunday child, a somewhat spoiled boy who has been given, or has taken upon himself, the task of carrying on the artistic traditions, cheered on by Mamma, flanked by Lisa and Greta, the two stay-at-home daughters, who together formed a matriarchy that was central to the family and that was based in a house in Ålsten from 1925 up until Greta's death in 1973.

Not a great deal is known about Gunnar's childhood. He was the third child; Pelle, who was born in 1875, was four years older and was the one who would take over Father's business; Lisa was two years older. The younger children arrived one by one in a steady stream.

Not all of them survived, but when the youngest sister, Signe, was born in 1893, the household consisted of four boys and five girls, and then there were the servants. At that stage they were living in a grand apartment at Norrtullsgatan 1, close to the gun-shop that during Gunnar's childhood years was just one of several stores that seem to have been a lucrative enterprise from the start.

The firm was old, no one knows exactly how old, but it was certainly fully active as early as 1742 and it has borne the name Widforss ever since Mauritz bought it in 1872. Not many people were licensed to deal in firearms at that time, and since Widforss was sufficiently forward-thinking to start mail-order sales there were opportunities for expansion. They also had shops on Drottninggatan, Vasagatan, Hamngatsbacken, and on the square at Hötorget, and they even had a branch in the Södermalm district. And behind the counter of one of these gun-shops I found a story that confirmed something Gunnar's relatives had told me about the family. I had heard from a number of sources that in its heyday the Widforss family was like something invented by Ingmar Bergman. The atmosphere, people said, closely resembled the film *Fanny and Alexander.*

I had no reason to doubt it, but, given that family legends have a tendency to take on the form of myths with the passing of time, I reserved judgement until by pure chance I learned that the author Sigfrid Siwertz (1882–1970) had been on the periphery of the family when he was a boy. His father had been employed as one of Widforss's shop assistants as a young man and eventually been promoted until he was the chief cashier of the shop on Vasagatan. The confirmation I needed was to be found in Sigfrid Siwertz's melancholy book of memoirs, *Being Young*.

> *My father's employer lived with his large family in a pleasant old house at the start of Norrtullsgatan, opposite Observatorieparken. At Christmas the house opened its arms to everyone. It could melt the frozen heart of the worst sourpuss. But, for me, miserable little wretch that I was, all of this business with the Christmas tree and the chain dance through all the rooms and the happy shrieks of excitement was so inexpressibly nerve-racking*

and sad that I went numb inside. Not even the little Russian peas I was usually so fond of with the Christmas ham were able to overcome my spleen. Everything seemed too bright and too much. I could not believe in it, could not bear it. I felt strangely alone and abandoned. I crept off into a corner ready to weep.

That's probably all we need to know about the environment that moulded the artist when he was young. Gunnar attended the Norra Latin Grammar School and then, when he was sixteen, began his training as a decorative painter at the Technical Institute. His summers were spent at Grisslehamn in the Stockholm Archipelago.

This last point is, of course, not without significance. No one becomes a landscape painter without having a childhood landscape. As an adult it is probably at least as difficult to learn the nuances of a crooked pine by the sea in the sun of high summer as it is to learn to speak a foreign language without an accent at the same age. During my researches I have seen watercolours of the archipelago that Gunnar painted during the war years that testify to a profound familiarity with the lapping of the water and the rocks in sheltered bays; smooth granite pavements, herring drifters and marine horizons that, in the hands of painters who have not seen these motifs a hundred times before they give a thought to painting them, can all too easily become banal. It may well be that I'm blind to faults that are close to home, and I am certainly much too loyal, but I like to think that Gunnar's keen eye for the colours and light of the islands, even for the scents, originates from the fact that he ran around there as a child before his ambitions were aroused.

. . .

Gunnar Widforss was five foot four inches tall. That is probably not really very significant, but I like to mention it if for no other reason than to add a touch of physicality to the picture of this 21-year-old, who, on a summer's day in 1901, after achieving distinction in his studies at the Technical Institute, registered his bicycle on the boat that was to take him via Hangö and Helsinki to St Petersburg and the first adventures of his journeyman travels. He was not big, not in any sense.

He remained in the Russian capital for a year and a half, employed by Erik Wagner, a master craftsman and one of many Swedes working in Russia at the time. It's apparent from letters home that Wagner was a family acquaintance, which did not prevent the relationship between master craftsman and journeyman sometimes being a stormy one. The job was poorly paid, and Gunnar immediately showed signs of the inability to handle money that reached great heights later in life. He was permanently broke. Wagner, moreover, seems to have been both a shrewd businessman and an emotionally unstable human being. Finally, in September 1902, Gunnar told him to go to hell, after which he did not even have enough money to take the tram and was forced to crawl to his father in one of his rare begging letters: he wrote to his mother about his dream of becoming an artist.

Ah, well, it all got sorted out and, for most of his time in Russia, Gunnar worked as well as he could and was not infrequently praised, even by the volatile

Wagner. He painted stairwells, shop signs and stoves, worked on the decoration of the foreign ministry and in the houses of directors whose parquet floors were polished twice a week. He rushed frantically from one client to the next, from one style to the next. It was everything from ceilings in Rococo, chapels in Moorish, transom lights in Greek, German Renaissance décor, sometimes floral painting and a good deal in a style covered by the general rubric Decadence.

In the beginning he had to live in Wagner's wallpaper store but eventually found a series of rented rooms in different parts of the city, all of which are described in detail in his letters to his mother. He drew complete town plans on which he marked his various addresses with the precision that was already becoming his hallmark. He learned Russian reasonably well and observed – wide-eyed with amazement, I like to think – the corruption and chaos of the metropolis. There was one street riot after another. One Saturday in March, the 'big revolution' took place on the Nevsky Prospect and there were many dead; the police had to clear it all up after the Cossacks received orders to draw their swords. Gunnar writes that it is always the students 'who cause the trouble' and his tone tells us which side he is on, but there are some clear signs of horror in his account of nocturnal disappearances and deportations to Siberia: 'Not even their relatives know where they have been taken.'

It is a tedious fact that Gunnar's correspondence never attains a very high stylistic level. Much consists of uninteresting small-talk about the weather and everyday matters – what he is eating, congratulations on special days and apologies for not having written more often. As to money problems, he will be writing about them for the rest of his life. Nevertheless these stacks of letters do offer – if no more than a glimpse here and there – good views of a reality wider than his private problems. The riots on the Nevsky rumble in the background. He can throw out a semi-ironic observation about factory workers in Russia only working 200 days a year because the other 165 are celebrated as holy days. On another occasion he uses his writing implement itself to make the point that this is a society on the verge of disintegration:

I'm writing with a pencil because I'm sitting in a restaurant having my dinner and it is forbidden to use ink because it seems to have dawned on the police that masses of undesirable business tends to be done in public houses.

...

In the late autumn his mother, Blenda, started asking about the price of fur coats in St Petersburg. Not unreasonable. She did, after all, come from a good family. And dear little Gunnar obligingly chased around checking the prices in shop windows, where prices proved to vary from 10,000 roubles for the very best down to around 40 roubles for the heaviest and coarsest. (To put that in perspective, Gunnar was earning 9 roubles a week.) His conclusion is that the fur coats for sale in pawnbrokers are the best value for money: it's possible to pick up a really good one for no more than 50 roubles. But he does wonder, all the same, whether it wouldn't be more worthwhile for Mother to buy a fur coat in Stockholm, since freight charges and customs duties are so expensive.

In his next letter, two weeks later, to his father

this time, an obviously surprised Gunnar thanks him for the fine sable fur coat. He would never have guessed that his mother was enquiring for him rather than for herself. He is so happy and grateful and he tells them, in the way only boys can, that he doesn't feel the cold at all now, not even when he is sitting on the roof of a tram in the December cold. Travelling up there is much more fun, in fact, and there's a lot more room.

. . .

By Christmas 1902 Gunnar was back home in Norrtullsgatan. His stay in St Petersburg had not, perhaps, been an unalloyed success, but he had learned a good deal and was now intending to do what his father wanted and set up in Stockholm with a view to gaining his master craftsman's certificate and becoming his own man in the security of a respectable bourgeois backwater. And Gunnar does indeed seem to have started and run a business for a while, but he failed to carry the project through, and at the start of 1904 he was on his way out into the world again.

This time he took the train south. His destination was Zurich in Switzerland and what he was supposed to do there remains rather unclear, but I get the impression that his main reason for travelling was to help him decide. To boost his confidence. As if he couldn't really bring himself to believe in his own abilities and had to flee from home not so much to see foreign countries as to find himself.

He wanted to improve his landscape painting, that much is clear, however, and he fantasized about long journeys on foot through Switzerland and southern France 'in order to see as much beauty as possible' – but reality turned out differently. After all, he couldn't keep on begging money from his father; he had to support himself; and with the approach of spring he set off for Geneva in the hope of finding a job. But he failed. Father sent him more money and Gunnar cringed and shrank to almost nothing. He was only happy on the days the sun was shining and he could paint beautiful views in the mountains. But when his mother wanted to show his watercolours to Georg von Rosen, a professor in the Academy, his reaction was one of fear and indecision: 'A painter that good will just laugh at them . . .'

She showed them anyway, and von Rosen found them worthy of words of encouragement, which she promptly passed on to her indecisive offspring, who was on the point of giving up all his dreams and following his father's advice to set himself up as a respectable master craftsman in Stockholm. He was twenty-four years old that summer. Blenda still hugged him to her like a child. There was little point in Mauritz saying what he thought.

Gunnar moved on to Munich, where he finally found work. Admittedly it wasn't of the same sad sort as in St Petersburg, but it was a job. The wages were small, but at least they were wages. He hung lining paper, painted façades, even safes, and was miserable, longed to go home, longed to get away, got the sack, wanted to be an artist or a scene-painter in the theatre in Vienna, perhaps design wallpaper patterns; and he dithered back and forth to such an extent that, having got thus far through the stack of letters, I wished with all my heart that he'd just go to blazes, which actually turned out to be Brixen in Austria, where he spent the autumn painting a monastic church in the Baroque style. Deadly boring work. He felt like a machine.

You've only yourself to blame, you chicken, I thought.

And when he finally succeeded in scraping together the fare to travel home and eat ham and peas and dance the chain dance at Christmas and possibly also to paint some timid little still lifes with his mother, I felt quite sure that, as ever, I'd been taken in by an out-and-out loser who was not only forgotten but deserved to be forgotten.

Chapter 5

The Involuntary Cocaine User

The gun-shop owner Mauritz Widforss went to join his ancestors in the autumn of 1905. Whether his death was expected or not is now hidden in the mists of the past, nor do we know how it affected Gunnar and his plans. After just a couple of weeks, however, Gunnar was off on his travels again, this time crossing the Atlantic to New York. We can assume that he was off to seek his fortune.

Exactly a hundred years later Johanna was steering the green Mustang with a sure touch northwards along the highway towards Mesquite, a flourishing small town in the desert in southern Nevada, bordering on the top-west corner of Arizona. It was a beautiful summer's day, and everything was peace and joy. We were travelling a little too fast for my taste, but any influence I have on issues of that kind is limited. Johanna was enjoying herself and mocking the Americans for being useless Sunday drivers with a spineless and incomprehensible respect for speed limits. Until, that is, she discovered that the speedometer was graduated in miles, not in kilometres, and we too fell back into the comfortable rhythm of the traffic on Interstate 15.

I sat with binoculars on my lap. There were a number of raptors in the air along the road and some of them – they looked black – were flying with their wings at an angle you don't see in any European species apart from the black-shouldered kite. What kind of bird was it? I thumbed through the bird book. Hardly an eagle. Buzzard perhaps; the size was right anyway. But none of the descriptions matched, and after a while I put the question aside, hoping to see the same bird again at a lower speed. I did, too, on every day of our journey.

I mention this just to underline my total ignorance of the avian fauna of North America at the start of our journey. If you don't know the turkey vulture, you really don't know anything. Everything is new, including the landscape. Mountains and light. Only the burger bar in Mesquite felt familiar. The thermometer read 94° Fahrenheit in the shade. We bought five litres of water and drove on.

Long before our departure from home I had decided to learn the names of birds as far as I could. Perhaps even butterflies and certain trees, but that's all. To try to understand the full range of the flora and fauna of a completely new and unfamiliar landscape in the course of couple of weeks is pointless, of course; just as well give up without even trying. The famous geology of the Colorado plateau was likewise included in my sphere of conscious disinterest, in spite of every book about the Grand Canyon containing at least one chapter on geology, frequently illustrated with a watercolour by Gunnar Widforss. It is difficult enough just to see all these strange rock formations and to digest the sights, but to learn the age and imaginative names of the sedimentary strata as well and to get them to stick in your memory would be as pointless as trying to photograph them without first contemplating them for months, perhaps years.

Birds, on the other hand, suit me as a provisional anchorage at least. Wherever you go, the birds of the area can be identified with the help of a bird book and a good set of binoculars. Not all of them, of course, but sufficiently many, especially for someone used to watching birds elsewhere, even if that elsewhere is on the other side of the world. I'm not claiming that this will provide the traveller with full knowledge within a few weeks, that certainly won't happen, but with the help of birds he will be able to spell his way through the polite phrases of the landscape in more or less the same way as a linguistically interested visitor to Japan learns to say 'Good day' and then to order a beer.

I learned to recognize the turkey vulture, to which Linnaeus gave the beautiful name *Cathartes aura*, a couple of hours later, between Rockville and Springdale on the threshold of the Zion National Park, where the scenery began to be so spectacularly beautiful as to render me speechless. At dusk the very same evening we drove up through the great forests towards Grand Canyon Lodge on the North Rim, where Gunnar went for the first time in July 1923 and wrote to his dear mother: 'I have never seen anything that can even approach this in majestic beauty!'

'Wonderfully, indescribably beautiful, magnificent!'

But by that time it was 1905 and he still described himself as a decorative painter in his passport. Unfortunately there are few letters from these first years in the United States, but we can nevertheless follow him at a distance both in thought and in terrain. He obviously had Swedish friends in New York, among them Percy Tham, who later became consul general and who helped him to find his feet. But the reasons why he made his way to Florida remain unknown – for that is where he ended up, in Jacksonville, as a poverty-stricken fence painter with dreams of being an artist.

He wasn't very interested in birds, as long as we exclude ostriches, which are, after all, something of a borderline case. At that time it seems to have been just about impossible to find a picture postcard in

Jacksonville that depicted anything else. Gunnar sent a card showing an ostrich on the racecourse complete with reins, sulky and driver. The American ostrich industry was experiencing a veritable boom in the years around 1900, and Jacksonville was the epicentre of the craze. I presume that the initial idea had been to produce the plumes demanded by the fashion designers of the day, but it wasn't long before the entertainment industry took over. The USA is always the USA. The whole town was like a funfair. Ostriches everywhere, and souvenir shops.

. . .

I saw an ostrich myself once, just one, from the train. It happened between Mombasa and Nairobi in the days I was travelling under a false name (Mr Swellengrebel) in the company of a girl who found it very easy to cry. It was a trait that, I'm afraid to say, wasn't improved by travelling with me.

Anyway, we had booked a sleeping compartment for two, and on that particular morning, before the girl – her name was Henny – started weeping, I was lying there looking out over the savannah that was passing by outside, when suddenly, right in the middle of everything, there was an enormous ostrich. It was over in a few seconds. This memory came back to me one day when I decided to check in my mother's drawer to see how much I had actually revealed to her in my letters home. It turned out to be quite a lot, but far from everything.

. . .

Apart from some time in St Augustine, said to be the oldest town in North America, where he spent his time with an Indian, an Austrian ex-officer and an American missionary, Gunnar stayed in Jacksonville for the whole of the first half of 1906. A lock-out directed at painters and decorators made his already difficult situation even worse. On top of it all he had toothache. And there were too many mosquitoes. Eventually he was given a commission – to decorate the bar in the Falstaff Restaurant. After this Mr King, a well-off sawmill owner, ordered an enormous painting, 10 x 6 ft, on the Italian model. The customer was satisfied. And the artist? 'I feel a complete humbug when I receive so much praise.'

A doctor wanted his dining room painted and an architect from Buffalo turned up with some other commissions. Things were on the move. Gunnar also sold one or two paintings at $5 apiece, but, as bad luck would have it, all his best sketches were stolen. He longed to go to California.

He moved to Brooklyn. Other Swedish painters and decorators encouraged him to go there, and there were jobs, that's for sure, but they were hard work and the pay was useless. He got work as a paperhanger, painted signs, failed when it came to portraits, copied masterpieces from photographs and sold them cheap to survive. If it hadn't been for games of cards with friends and the entertainments at Coney Island, he wouldn't have lasted a day. Brooklyn was swarming with Swedes, several thousand of them, particularly from Småland and Öland, most of them Pietists, among whom there were girls on the lookout for a husband and they had a weakness for educated Stockholmers. You had to be on your guard.

'It would be too bad if something like that were to happen; it would mean the end of all my plans to really make something of myself eventually.'

One of the people for whom Gunnar painted signs was a Dr Borgström, a Swedish dentist in Manhattan, and, since our hero was still plagued by toothache, they developed a joint scheme. He gave a fleeting account of this business deal to his mother Blenda in a long and well-written letter in English, written perhaps as language practice. It was the late winter of 1907. An important letter, I think, which in parts expresses a fear of getting old, or of already being old: 'I am able to learn all kinds of paintings. I only fear it is too late now. I am getting old, not in views or thoughts, but of age and that is bad enough for me.' It's as if some things are more easily expressed in a foreign language.

The man holding the pen was twenty-seven years old. No age, we might think, but his dealings with his fellow-countryman in Manhattan made it easy to understand his melancholy frame of mind: the dentist had extracted virtually all his teeth and then set about restructuring and rebuilding. 'Bridgework do they call it,' Gunnar wrote and made the operation sound like one of the series of brilliant innovations that was making America a land of the future. The only thing he really complained about was the cocaine, the anaesthetic. It gave him such a dreadful headache.

The payment, too, was a worry that troubled him for months. The upshot was that Gunnar got new teeth in exchange for painting a portrait of Borgström's wife – and he couldn't paint portraits. They simply did not work. And he knew it, but it would be several years before he finally gave up trying. Nothing works in these attempts. The models look as if they are stuffed, and he was mortified and ashamed: 'The fact is that my portrait painting is a big humbug.'

Of the few portraits I have seen, there are only three that really amount to much, and two of them – one watercolour and one in oils – not unsurprisingly depict Blenda. All the rest fail, including the self-portraits. With time, people will simply vanish from his paintings.

. . .

Gunnar Widforss is today completely forgotten at home. In America he is honoured as one of the great painters in his genre, whereas in Sweden even the most discerning connoisseurs have question marks written all over their faces when you mention his name. There are a number of explanations for this, one of them being, I believe, something as simple as everyday bad luck. It could all have been so different. In July 1923, for example, if Gunnar's first visit to the northern rim of the Grand Canyon had taken place just two weeks earlier than it did, he would have met a man who had a rare influence on the cultural life of Sweden, one of the influential members of the Swedish Academy and, moreover, a talented watercolourist who would probably have recognized the quality of Gunnar's work. From personal experience this man knew the difficulties and wrote about the impossibility of painting in this very place.

> *For these colours are magnificent, but they are light and discreet, not intense or glaring; their effect is dreamy, rose-garden, empire style, the pink, violet or pale green*

dresses worn by young girls in a ballroom; they have the feel of clouds lit by the red glow of morning, and because of them the whole of this landscape seems so light and airy that it looks as if it could be blown away by the first puff of wind.

The explorer and travel writer Sven Hedin, North Rim, 4 July 1923. A little piece of happy rose-scented whimsy at the end of his book about the national park and obviously composed after he had managed to forget the words he wrote in the first chapter: 'a poet would only make himself ridiculous if he tried to sing of the Grand Canyon.' It's not one of Hedin's better books. There is too much geology. And his favourite travelling companion seems to have been a thermometer. Exciting for a meteorologist perhaps, but of little interest otherwise, we might think, even to the great author's mother.

Because that is who he was writing for. There is a sense in which he was always doing that, but his book about the Grand Canyon, which was published in 1925, is literally a series of lightly edited letters to his 'guardian angel' back at home on Blasieholmen in Stockholm. 'Not a day was allowed to end without a few pages being written to Mamma. There is no one who has followed my adventures with a keener and warmer interest than she has done. In her thoughts and in her prayers she has always been beside me along the lone and desolate roads.'

It is an irony of fate that these two mummy's boys never had the chance to meet there, just there, just then, and feel the same dizziness in the dusk at Bright Angel Point. Sven Hedin would have loved the watercolours Gunnar had painted in the early summer in Zion and in Bryce Canyon. They were lying in the trunk of his car and very soon they would bring him his well-deserved breakthrough in Washington, D.C. But we mustn't get ahead of ourselves.

Chapter 6

With His Sights on the Paris Salon

The difficulty lies in coming to a decision. There are, of course, plenty of other obstacles and difficulties along the way, there always are, but they rarely hinder us as much as the great life-defining decisions, by which I mean the kind of decisions that cannot be made without simultaneously sealing off all avenues of retreat. No way out, not even a hint of one. Becoming a good painter is hard enough in itself, but to make a decision once and for all and then not yield an inch is actually even more difficult. It is not until the artist lets go that his vision changes. It's as if the ultimate precondition for success in painting is the muffled echo of the fear you will make a complete mess of it.

That is why I never became a good photographer. Everything was in place, but what was missing was the decision. The rock-solid promise. I left the photography course after my first term just as lost as when I'd started.

The idea had seemed sensible enough, indeed more than sensible. I was to follow in my father's footsteps and become a photographer. Ansel Adams was my great idol, along with Andreas Feininger, the man whose books Dad had translated into Swedish even before I was born. For as long as I could remember, I'd had all these pictures around me: Henri Cartier-Bresson, Edward Weston, Christer Strömholm, Brassaï, all the great names. And for the whole of my childhood all the *US Camera Annuals* from the 1940s were lined up on the bottom shelf but one of the bookcase, within easy reach of the smallest child. Even today I still know them inside out. The visual language of black-and-white photographs had been mine since birth, the darkroom just as natural a part of the house as the bathroom and the boiler room. I was six years old when I got my first camera, twenty when I started at photography school – still too young and far too indecisive when faced with all the possibilities of those years.

My inglorious exit from the visual arts might have been the result of my lack of talent, an absence of the photographer's eye, but I would prefer to ascribe it to indecision. Later in life I've reached decisions about all sorts of things, and I think I know how vital it is sometimes to just let things go and put up with the fear of failure.

I would like to think that making a decision was what saved Gunnar Widforss. Something happened, no one knows what, but it must have been shortly after his return home following his disappointments in Brooklyn. His two years on the other side of the ocean hadn't really given him much more than a fine set of teeth, but in Stockholm, at some point in 1908, Gunnar finally made up his mind to become what he wanted to be, or, rather, to become who he wanted to be. A watercolourist. An artist. Free.

The following year, when he set off south again at

the start of six years' travelling around Southern and Central Europe, his passport described him as an artist, and from this point on there is an air of confidence in his letters, or of decisiveness at least. His naivety, perhaps we should say foolishness, is the same as before, however: it led him to head straight for Monte Carlo, and, once the hardest step had finally been taken, to go to the casino and lose all his travelling funds. Back to zero, as usual. What an idiot! It's easy to see why he could never have made it as a businessman. Poor fool. Thirty years old and broke in Marseilles.

At this point I recognize that nothing would be simpler than to construct a pretty conventional myth of the artist around Gunnar Widforss, beautiful and half true, as such romantic legends often are. It's tempting: there is a tradition of frescoes painted in the red of blood and the green of absinthe in honour of artistic struggle and blind despair in distant lands. But I'm not too sure about it, and there is something that doesn't fit. Early Widforss reminds me to a worrying extent of late Kallstenius. He produced many – too many – paintings I could easily have walked past without bothering about, had it not been for the signature and for my sense of loyalty.

What we can say about Gunnar's watercolours, however, is that they became markedly better around the start of the 1910s. Before that he could pull it off on occasion, but, as far as colour, line and composition were concerned, it was still too easy to mistake him for other painters, for amateur artists even. But practice began to pay off. Slowly but surely he made the colours his own and he painted with the sort of clear, shimmering lightness that, in combination with his typical choice of subject, means that even a relatively untrained eye can recognize from afar a Widforss watercolour from his great period. No one else painted the air as he did, for instance. When he was by the Mediterranean, he practised painting the air with great single-mindedness.

I have to admit that I have seen no more than a fraction of what he produced during his European travels before the First World War. For reasons I don't understand, remarkably few of the hundreds of watercolours – possibly up to a thousand – that he sold or gave away ever came on to the market, and those that did usually ended up in the United States, which is where the collectors are. In spite of that, I believe I can trace his development, influenced as I am by what he himself wrote in his letters to his mother: he wrote of progress, working routine and the low spirits that can be cured only by beautiful subjects. There was nothing in his imagination that was worth painting, and he had done more than enough copying. It simply wouldn't do. He had to have his subject right in front of his eyes, in exactly the right light, otherwise nothing worked. The weather, the light and his mood had to last for whatever number of days it took for the painting to become good. Simpler pieces, as he called them, could be painted very quickly and sold on the spot to provide him with a living. In a letter from Martigues on the French Riviera, in the summer of 1909, he tells of an occasion when he painted seven watercolours in three hours. Trivial pieces he pressed on cafés to keep hunger at bay. This was just after he had been defrauded by a German painter who was supposed to be looking after his business affairs in Nice: the German had pawned fifty of Gunnar's better canvases for such a considerable sum – which he pocketed himself – that Gunnar could never afford to redeem them.

But he thrived on the Riviera, thanks to the heat and the views and perhaps even to the travelling itself. Only exceptionally did he stay anywhere more than a couple of months, and for the most part he drifted from place to place by the sea or in the mountains. Cannes, Antibes, La Turbie, Menton, Venice, Capri, Chamonix, Geneva, Merano, Thun, Wengen, Lugano, Arco: he moved from one to another and he painted. I attempt to read between the lines of his letters to see whether he was living the wild life and generally behaving as artists are expected to, and no doubt he did now and again, but for the most part he simply seems to have been working hard. When he does mention something exciting, it is usually only a passing reference before he returns to droning on about the weather and his meagre finances.

Thus, for example, he writes from Capri, where he lived for a couple of months in 1912 'in a very artistic environment' with a mysterious Indian whose wife was an English artist: 'I have also made the acquaintance of Maxim Gorki, who has lived here since the Russian revolution, and of a good many other Russians, mostly writers.' And then nothing more. Among the scenes from Capri listed in the catalogue of an exhibition the following year, we discover that he visited San Michele, though neither he nor the archives make any mention of possible contact with Axel Munthe.

In fact few names are mentioned in his letters. Apart from the fraudster in Nice, he associated with a couple of other German artists at various points – Georges Einbeck and Otto Pippel, both of whom later achieved a modicum of fame – and in Menton he was praised and given sound advice by the ancient Frenchman Henri Joseph Harpignies (1819–1916), the master watercolourist and one of the leading figures of the Barbizon school. Generally speaking, however, he did not encounter many people with names famous enough to tempt the present author out along promising by-ways. He frequently complains of loneliness and it's obvious that he lacked a network, something just as important in the art world then as it is now. Was he just diffident, perhaps?

He did meet King Gustav V, however, who was

staying at Cap Martin outside Monaco. Gunnar was given an audience in both 1910 and 1912 and sold no fewer than six watercolours to the 'very aimable' monarch, though, unfortunately, all but one have long since disappeared. And he considered himself talented enough to risk applying to the Salon de Paris in 1912, at that time by far the most important art event in Europe. An artist who exhibited at the Salon was someone to be reckoned with. A man of the future. Gunnar sent in two watercolours, and amazingly both were accepted. His ambitions immediately moved up a notch, and he hoped for a breakthrough in the press. Nothing came of it, of course. He was living in absolutely the wrong age. Naturalistic landscape painting was not something likely to cause a stir in the year that visitors to the Salon were flocking around *The Enigma of the Oracle* by the up-and-coming Giorgio de Chirico. Fate decreed that the Modernists would wipe the floor with Gunnar Widforss.

. . .

Let me spell out a couple of things at this point. I am not an expert in anything apart from odd bits and pieces of no relevance in the present context. Largely for my own greater pleasure I have acquired a knowledge of art that goes a bit beyond what usually passes as general education, but I am not an art historian and I am certainly not versed in that rather amorphous discipline called theory of art.

So I don't have a particular axe to grind. I do find it rather a pity, though, that the art establishment has branded the kind of beauty that is still alive and well in literature and music as utterly antiquated in the visual arts, which is not to say that I feel any great urge to grumble about the snares inherent in Modernist freedom or the pitfalls of Postmodernism. As to the excesses of the present day, I am for the most part astonished but remain essentially open-minded for all that. If, like Marco Evaristti, you want to slaughter goldfish in a blender, or like Bob Flanagan at the New Museum of Contemporary Art in New York, you want to nail your own genitals to a plank with an ordinary galvanized nail in front of a devoted following, I won't be the one who calls the police. It's just that I don't like it.

The idea we often hear voiced – that current performance and installation artists are difficult to understand – is not a view I share. During the last couple of centuries art has rarely been easier to understand than it is at the moment. Iconoclastic social criticism and provocation. The alternation of disgust and cheerful irony. A child can understand what is being aimed at, particularly since the vaguely philosophical thoughts being put forward are usually of the kind that have come to be termed cod philosophy. And with good reason. Plain and simple rubbish. Nothing to make a fuss about. Anyway, I'm more of a biologist than a connoisseur of art and consequently I'm mainly interested in sex. As a driving force, I mean.

. . .

Many of the places in which Gunnar worked and lived before the First World War were health spas. I don't suppose it was a conscious policy on his part, not in the beginning anyway, but it's just how things fell out. It was partly that sanatoria and the smartest resort hotels were located where the views were best, and partly because these spas provided a good supply of potential customers, affluent and in love and sometimes frail enough to be anticipating the end, with all that can imply for people's perception of landscape and art.

Hotels were not cheap places to live for a man with Gunnar's earnings, but they undoubtedly paid off because they brought him into contact with fellow-guests who were happy to buy his paintings. During most of his early travels he had lived in unpretentious rented rooms, usually alone, but from now on he stayed more and more frequently in hotels where, for most residents, time is in some sense rigidly demarcated. For most of his life, apart from his parents' home in Stockholm, Gunnar never had a real home, and consequently, like a refugee, he did not own more than would fit into a suitcase.

I can't be sure, but I do believe that a change of air – that long-since discredited panacea that doctors prescribed for well-heeled Europeans before the wars and before penicillin – worked amazingly well precisely because travelling to the seaside or to the mountains broke the glum monotony of daily routine. It also encouraged the gently narcotic melancholy induced by constant leave-taking and the quiet eroticism that can soothe some of the ailments that have names that grow longer with the passing years while essentially meaning the same thing – colitis, neurasthenia, chronic fatigue syndrome.

Gunnar, of course, fell in love with more than just landscapes, but there are limits to what a mother can be told, however cherished she may be. He was nevertheless fairly open with her. A Miss Isberg, a gymnastics instructor, turned up from somewhere and was present in his life for a few years, initially as an acquaintance, then as a friend. He wanted to get married. He wrote to Mamma and asked for help with his birth and residence certificate, which lacked the required information about his stay in the USA and thus made it impossible for him to get married unless he put a notice in the American papers to ascertain whether there were women over there with a claim on his person. But Miss Isberg's interest proved to be cool and nothing came of it. And anyway, Gunnar couldn't afford it.

His attachment to Carl Erik Häggart (1880–1940), a pharmacist and old schoolfriend who suffered rather poor health in his youth and consequently spent his winters in Menton, was very much stronger. Their paths crossed in March 1910, after which they corresponded up until Gunnar's death. If their letters have survived, I have no idea where they are, but the close links between the two are obvious from the way Gunnar wrote of Häggart in his letters home. I don't think anyone was closer to him. Häggart was the friend Gunnar needed, a support in matters big and small, always there for him, able to lend him money when his purse was empty and to act as his agent in dealings with gallery owners and newspaper editors. And when the end came, it was Carl Erik Häggart who wrote Gunnar's obituary in *Svenska Dagbladet*. We shall have reason to return to him.

For a clearer understanding of Gunnar and his career, however, we have to remember that his first solo exhibition, in Stockholm in January 1913, was not a great success, though it wasn't a complete failure either. About half of the fifty watercolours from Capri and the Riviera that were shown at Hultberg's art shop found buyers, and they brought in sufficient funds to finance studies in Paris that spring. Where he studied and how much he studied is uncertain, partly because Blenda accompanied him in order to supervise his progress. Which means there were no letters. (There are times when the woman gets on your nerves.)

My guess is that his studies were not particularly well organized but were sufficient for Gunnar to be able to claim – with honour intact – to have studied in Paris. This would later impress the Americans: he would tell them how the proudest moment of his life as an artist came during that spring in Paris, when Anders Zorn not only expressed his approval of Gunnar's watercolours but went so far as to buy one. His self-confidence bloomed. His journeyman days were behind him. He accompanied his mother home from Paris and set about putting together another exhibition – for December – of scenes from Stockholm and Grisslehamn. Stockholm City Hall bought two of his watercolours.

. . .

I rang the buildings manager at City Hall and told him I'd read an old article in the *Los Angeles Times* that stated there were two watercolours by Gunnar Widforss, both signed 1913, hanging in City Hall. Somewhere. I was really keen to see what little there was to see. Gunnar had never been particularly well represented in galleries in Europe, although as early as 1912 he was convinced that he would end up in the National Museum sooner or later. And he did, even though it was posthumously, when the museum acquired the large painting of the Grand Canyon that the head of the US National Park Service had once presented to Gustav Adolf, the then crown prince of Sweden. Apart from that, an art gallery in Dresden has a 1912 watercolour, the city museum in Stockholm owns one painting, as does the Royal Household. The rest are in the United States, including the Smithsonian American Art Museum in Washington and the Museum of Northern Arizona in Flagstaff. But back to the City Hall.

'Did you say Widforss?'

'Yes, exactly – with a *w* at the start and double *s* at the end.'

'Give me a moment while I check the inventory.'

After a short while the buildings manager picked up the receiver again. I had a bite! There were two watercolours listed in the inventory and they had indeed been acquired direct from the artist in 1913. There was a larger one, *Skeppsbron and Söder*, and a slightly smaller one, *Stadsgårdshamnen and Söder from the Sea*. Their combined value was stated as 250 kronor, that being the sum paid for them.

'May I come and see them?'

'Of course. But let me find out which rooms they are hanging in.'

He promised to get back to me, which he did, six months later and after some prodding. Unfortunately he had failed to locate the paintings. The City Hall is so enormous, you understand, but he hadn't given up hope. But I had. In all probability the paintings have been stolen, possibly decades ago. No one knows. But they are gone, that's for sure. My guess is that they will be hanging in the home of a rich private collector in the United States.

It's never nice when artworks are stolen. But in

some cases it's possible to comfort yourself with the idea that the artist must be smiling in heaven at the thought that his work is now so highly valued and desirable that real gangsters with real machine guns and the whole caboodle find it worthwhile carrying out raids that are starting to resemble real bank raids. Rembrandt, Munch, Renoir and Strindberg. I'm quite convinced that the last would have approved of the raid on the Strindberg Museum on Drottninggatan, not least because of the valuation of tens of millions that was then put on *Night of Jealousy*, a painting that no one understood and certainly wouldn't have paid good money for when it was painted in the 1890s. But things stolen silently, without even the owner noticing, those are the things that make you weep.

Chapter 7

Aimless Journeys

The last place in southern Utah is called Jacob Lake. There's a gas station there. If you then follow the road south, forty-five miles through a forest called the Kaibab Plateau, you will eventually come to Grand Canyon Lodge, a legendary hotel in the American wilderness Baroque style, built in 1928 of undressed logs and rough stones and sort of hanging on the edge of the abyss.

A strange journey, particularly for anyone coming up from the desert the same day. And I mean up, because it is the height above sea-level that defines the visitor's impressions more than anything else. At Jacob

Lake you are already a little over 7,900 feet, and the road south from there runs gently uphill. When we drove that way at the start of June, there were still snowdrifts lying in the north-facing edge of the forest. There is so much snow every winter that the road is closed between late October and early May.

I've no wish to give an account of our wanderings on the Colorado plateau. The car was fine, as were the driver and the roads, and we covered several thousand miles through four states. Hither and thither, back and forth. Current form, the weather and our mood decided things for us. The United States is made for that sort of unplanned car travel. The only thing we had decided in advance was that we would drive up to the North Rim on the first day, spend a couple of nights there and then carry on somewhere else, anywhere else, in fact, in order to give us the chance to return to the same place a couple of weeks later. There is nothing quite like returning to a place.

So we spent only a short time just inside the park boundary at a spot where there was a large sign at the side of the road announcing that here, right here, was the start of what was by far the most beautiful hiking trail in the whole of the Grand Canyon: it's called the Widforss Trail and it leads for about six miles through the ancient forest along the northern rim of the canyon to Widforss Point. It seemed to me a touch over-hasty to go there straight away. We would return.

Nor do I want to describe the Grand Canyon. I can't anyway. Gunnar could, and before him the famous romantic Thomas Moran could, but not many others. This much I can say, however: it's big, it opens up suddenly without any prior warning, it's absolutely vertical and so awesome that anyone who lacks the ability to fall into ecstatic raptures of the evangelical born-again variety cannot truly profit from the experience. It's simply too much. That is probably why most descriptions drown in detail about the age and variable weathering revealed by the different geological formations. You feel the need to say something. You can't see the Colorado River, nor can you hear it as it flows along 3,000 feet below your feet. Even in the 1920s nine out of ten visitors only went to the South Rim, more than seven miles as the crow flies from Grand

Canyon Lodge on the opposite side of the abyss. It's possible to take your life in your hands and hike down trails to the river, cross the bridge and climb up the other side, but it takes several days and the heat is ferocious. Drive round by car and the distance is about 220 miles. So the North Rim and the South Rim are different worlds. The South Rim was exploited first: El Tovar Hotel was built in 1905 and the Santa Fe Railroad had already reached that point. And the South Rim lies at a lower altitude so the climate is warmer and drier, the forest cover sparser and bushier, more or less like southern Europe. Everything is open all year round. Millions of tourists.

We went on a beautiful day. Gunnar is buried in the churchyard there. But it is the clear air and the dizzying feel of the higher side, of the North Rim, that I want to talk about. But how? We walked to Bright Angel Point in the twilight. It's beyond all compare – the words simply don't exist. My only comfort was the thought of returning there. And then, quite suddenly, a member of a small group of Polish tourists could no longer control his emotions and he broke into a melancholy song in the language of his homeland. He sang out into space, eyes closed, voice firm and strong. Some Japanese, cameras on auto-focus, moved away with brisk little steps, clearly disconcerted, while the Americans smiled in understanding.

It is not easy to give an account of all Gunnar's travels, nor perhaps is it necessary. It's just that it tends to be conventional in the case of artists' biographies. In standard necrologues the career of the deceased is laid out in the form of a trellis of posts and appointments up which the lies can climb, but for painters it is their travels that matter. It might sometimes be appropriate, but more often than not it seems to me that these travel narratives are simply confusing. And, in Gunnar's case, ultimately a mystery.

So he set off again in early 1914, back to Arco in the South Tyrol, to Lake Garda and to Vienna. With the approach of war he was living as temporary guest in an old castle in Brogyán, a village that is called something different these days and that is now in Slovakia rather than Hungary. He had been invited there by Natalia, the widow of Duke Elimar of

Oldenburg, who was interested in art and had invited him after seeing his watercolours in Arco in the spring. An older woman in a beautiful castle in picturesque surroundings. As stylish a conclusion to his European travels as anyone could ask for.

Then came his Scandinavian travels. A whole generation of artists has testified to the isolation caused by the war and their dreary wait for light and community. Part of it, as always, is a post-hoc invention, but there is no doubt that many artists felt cut off from the circles on the Continent, where the surrealist nightmare of the trenches was contributing to the demolition of the last barriers holding back the floodtide of Modernism. One 'ism' followed on the heels of another in rapid succession, and the only thing that mattered was keeping abreast of the changes. Not that it concerned Gunnar Widforss; he had found his style and his subject matter, which existed everywhere – including Scandinavia – even if the light there was not the light he esteemed most highly.

During the war years Gunnar travelled far and wide around Sweden, Norway and Denmark, constantly in search of new scenes and townscapes to paint. He spent a lot of time in the mountains, in Åre and Storlien and even further north in Abisko and in the Lofotens. He lived for a time in Visby on Gotland, but also on the west coast in Uddevalla and Lysekil. Helsingör, Trondheim, Ulvön, Sala, everywhere. And, of course, summers were spent in the Stockholm archipelago – Utö and Grisslehamn. I may be wrong, but I get the impression that he was managing well, thriving and earning money.

What is absolutely beyond question, however, is that his exhibition at Hultberg's art gallery in November 1917 was his most successful exhibition ever. Even in the early 1930s his letters are still referring back to this success. Out of roughly fifty watercolours thirty-seven were sold in two weeks and earned him a sum of 11,400 kronor. For him that was a fortune. The statement of accounts still survives, as does the long list, handwritten by the artist himself, of friends and others invited to the opening. The *Roskär Pine*, the painting I had failed to acquire, was probably one of the paintings sold at this time, and if the others were of the same quality the success of the exhibition is understandable, as is the self-confidence revealed by Gunnar having the courage personally to invite Gottfrid Kallstenius, Anders Zorn, J. A. G. Acke, Ernst Thiel, Oskar Bergman, Georg von Rosen, Axel Sjöberg, Carl Milles, Carl Larsson, Bruno Liljefors, Nils Kreuger, Karl Nordström, Sven Hedin, Eugene Jansson, Olle Hjortzberg, Carl Eldh, Albert Engström, Anton Genberg, Gustaf Fjæstad and the rest of the artistic jet set of the time. (The list includes nearly 300 names.)

The way ahead was clear. Gunnar was thirty-eight years old and had begun to make a name. He had long been competent but had not had enough of a name to live well on his art. Now it was the shop, his father's old gun-shop, that borrowed money from him rather than the other way round. It is said in the family that he courted a girl, but she wasn't interested. Perhaps that's why he went to North Africa once the war was over. How should I know?

Tunisia. Right down into the desert. Alone. Why?

In a letter to his mother he writes:

> *I wish I was a savage and had never seen a painting and then, perhaps, I would be able to appreciate the colours better or more exactly. The ideas about light and shade I have had up to now are quite impossible to implement here … The shadows are so full of light – it's a pure pleasure to see.*

Everything was new. 'You can see great flocks of some kind of white seabird. Don't know what they are – big as swans.' Someone who doesn't recognize pelicans really doesn't know very much. And the only thing anyone reading his letters home to his mother would be aware of is that Gunnar wanted to paint things that were different from the things he was already master of, all with a view to another exhibition in Stockholm in the spring. It proved difficult, the light was too sharp, but he stuck at it and he got a lot done in a couple of months in distant oases down close to the Algerian border, but also in Sousse, Tunis and Cairo, where the sight of him and his easel caused such a stir that he had to hire a boy at four francs a day to organize a queuing system for all the nosy youngsters in the district.

Returning to Provence in January 1920, Gunnar had a whole load of paintings ready to exhibit. Houses and lanes, palm trees, camels, desert scenes. We don't know what happened, but, since some of these African scenes are still in the family, there is reason to believe that the public was less enthusiastic this time. The exhibition was not due to take place until April anyway, and to fill the time our hero roamed around southern Europe like a lost soul. Martigues, Saint-Raphaël, Menton, Castillon, Sainte-Agnès, Grasse, Nice, Ventimiglia, Genoa, Milan, Como, Basle. By this stage of my reading of his letters, I had begun to suspect that something was wrong. Mamma was not getting the whole truth. After Como in March there was a break in the correspondence. The next postcard was written in New York, in January 1921. I decided to change tack and follow a different line.

. . .

There is much that we never tell anyone. Looking back on my own journey around the world many years ago, I didn't write a word to my parents about my reasons, about what actually spurred me on – apart from the self-deceptions, that is, the claims to be filled with a yearning for adventure of the kind that deep down inside I had always wanted to avoid. But there were also numerous significant events and many concrete experiences that simply did not figure in my letters.

I told them I was ill, but not a lot more than that. It was only much later that I told them how the doctors in the Prince of Wales Hospital in Sydney queued up outside my isolation room to get the chance to see and to prod this strange case that might be the first AIDS patient in Australia. I think they were actually hoping I was. It was 1981. AIDS did not exist as a concept then, and the illness had a different name, the virus being called HTLV-III, if I remember rightly. I did not understand anything about it. I was happy with all the attention and to be looked after.

But the questions worried me. The same questions time after time. They kept at it for two days and nights, a number of different doctors keeping up a long interrogation, trying to coax out of me what I'd been up to in Central Africa six months before. About sex. I told them the truth, but they weren't going to give up that easily. They thought I was hiding something. So they continued, one after the other, explaining in their gentlest voices how important it was that I told the truth. That there was nothing to be ashamed of. That I didn't need to be afraid.

Not until the answers came back from the lab, at the same time as I went as yellow as a Swedish postman's bike, did they give up and disappear. A false alarm. But at least I got to keep the room.

Chapter 8

The Man on Saint Martin

I read all the letters in a single go. A life. It took several weeks and it gave me a good picture, but a one-sided one. Consequently, quite early on I began to dream of other sources. I had quickly recognized that Carl Erik Häggart, the chemist Gunnar had met in Menton, was his closest friend, and bit by bit I began to follow a trail that led quite unexpectedly into the heart of the Swedish pharmaceutical industry. It was Gunnar himself who provided me with the first clue. After a couple of years in the United States he wrote to his mother: 'And Carl Erik is now managing director of Astra. Hope things at home are in a good enough state for him to be able to put the company in good shape. If anyone can do it he can.'

AstraZeneca. Today it is one of the major pharmaceutical companies on the global stage. It would not be unreasonable, I thought, to expect there to be a bottomless company archive with information and papers of every kind about the man whose collected correspondence I am seeking. If such correspondence exists, that is.

The staff in the information section had obviously attended every available course on the art of dealing with strange questions with good humour and in the cheerful kind of tone that gave you to understand that there was nothing they would like more than to deal with your request. They admitted they had never heard of Häggart, but it took no more than a few seconds for the basic facts to be on the table. Astra, when all was said and done, had only had nine managing directors from its formation in 1913 to its merger with Zeneca in 1999. Four before Häggart and four after. He held the post from 1922 to 1927, and they turned out to have been dark and gloomy years in the history of the company. He owned a seventh of the shares himself but was forced out of his post and from 1930 on he was managing director of Pharmacia – their rival – instead.

Häggart had been with Astra from the start, both during its early happy days and during the fatally bad wartime speculations that led to the company being taken over by the Wine and Spirits Monopoly, i.e., the state. That just happened to coincide with the few months in 1920 when the country had its first Social Democratic government, which was led by Hjalmar Branting, who just happened to be a neighbour of the Widforss family on Norrtullsgatan. There were plans to nationalize the whole of the pharmaceutical industry, but the government fell before anything came of it, after which the aim became to minimize the losses by finding a buyer. Thus, for a while, Carl Erik Häggart found himself negotiating with Ivan Bratt, chairman of Astra and managing director of Wine and Spirits, to buy the whole company himself. He actually ended up with just one seventh of it.

It seems to have been a dynamic period in every

sense. A retired chief information officer I was put in contact with told me, for example, that, long before the company began making money on ski-wax and medication for gastric ulcers, Astra had manufactured various kinds of ointments that consisted largely of zinc oxide. Laying your hands on a supply of zinc was easier said than done, especially during the First World War. As politely as possible I tried to bring the conversation round to Häggart, but the enthusiastic pensioner rapidly got into his stride and proceeded to tell me how a colleague of quite exemplary ingenuity had succeeded in circumventing the problematic shortage of zinc by buying and dismantling the roof of a whole manor house in Skåne and dissolving the sheets of zinc in sulphuric acid.

'What about that, then?' he said.

I managed to elicit from him a hint that occasional stories about Häggart had circulated, though they were pretty trivial. The short character sketch printed in a lavish book on the history of the company was presumably written with his posthumous reputation in mind: 'He was very popular, particularly among

Astra's ladies.' He never married, however, the old information officer was quite certain of that, and he had left no traces in the company archives, no traces of a more personal kind anyway.

'Children?'

I asked the question rather hastily, since the man I was talking to belonged to my parents' generation, born in the 1920s and consequently possibly inhibited by Victorian conventions as to what could or could not be said about children who at that time would have been called illegitimate. But he didn't say anything. Hummed and hawed for a while and then pointed me in the direction of Pharmacia, where they possibly knew more. I realized that he knew something but had no intention of telling me. A hint was all I was given and a pointer that proved worthless, since Pharmacia had no knowledge at all of its own history. It seemed that everything had been jettisoned when the company was sold to Pfizer a few years before. As far as they were concerned Häggart had been consigned even more deeply into oblivion, if that's possible.

It was as if the ground had opened and swallowed the man. I was getting nowhere and I eventually gave up. Instead, I began snooping rather half-heartedly through the relations of another of Gunnar's friends, who was an amateur painter and the director of a chocolate factory during the 1920s in San Francisco.

Time passed. The Häggart trail had gone cold, icy cold, but both the man and his significance often came up in conversations with Gunnar's nieces and nephews, particularly after a well-written and moving letter of condolence came to light; it was dated December 1934 and it was addressed to 'Dear Aunt Blenda' from 'Yours affectionately' Carl Erik. It offered the sort of comfort that friends are there to offer.

Another piece of the puzzle was added when I came into possession of a couple of telephone numbers in connection with a man I knew nothing about, absolutely nothing apart from that he might possibly be a relation of Häggart. A wild chance, a shot in the dark, but I was sufficiently curious to ring the numbers. No answer. I rang again, many times, over a lengthy period. No answer. And then suddenly one day there came a curt answer from a man at the other end. It was a mobile number. I was thrown off balance but quickly pulled myself together and explained what I was up to. That I was following up Carl Erik Häggart, the director of pharmaceutical companies.

'He was my grandfather. On my father's side. Why are you interested in him?'

'It's a long story. Am I ringing at an inconvenient time?' I asked.

'Well, it is really. The fact is I'm on Saint Martin – the island in the Caribbean. Cruising. I forgot to turn my mobile off.'

That threw me again. Completely. What could I say?

'It's actually Gunnar Widforss I'm enquiring about. The watercolourist. They corresponded.'

'Yes, I know,' said the man on Saint Martin. 'I have all the letters at home. Paintings, too. A stack of newspaper cuttings and photos. If it's Widforss you're after, I'm sitting on a goldmine. Give me a ring in a week. I'll be home again by then.'

It was a strange feeling, though not altogether new.

I'd had a similar experience not very long before, but on that occasion I was the one who had forgotten to switch off his mobile.

. . .

'Is this a convenient time to talk?' a woman's voice asked.

'No, definitely not. I'm in the car. Sitting at a red light,' I said, peering out through the side window. 'At the corner of Paradise Road and Las Vegas Boulevard.'

'Oh!'

'It's all right,' I said.

All right was scarcely the right phrase to describe the chaos a visitor is thrown into on the streets of Las Vegas. But the conversation came like a cool breeze, for I knew that the woman on the other end of the line was also on the other side of the world in a landscape of divine beauty in which the blue of dusk would just be beginning to colour the late spring evening. About as far from the midday heat of the Nevada desert as it is possible to get. A call from the moon would, by comparison, have seemed like a local call.

The light changed to green. She gave me a brief and hasty summary, which is what we usually do over long distances. She wondered whether I would consider giving an address in memory of her father, my friend Gunnar Brusewitz, who had died almost a year before. It involved the dedication of a building that was to be named after him.

I don't usually agree to anything before sleeping on it, because I'm so easily swayed by flattery that I get carried away. But on this occasion I was in no doubt and so I made an exception. I was convinced that Gunnar would have enjoyed a good laugh at the thought of me sitting in a Mustang on Las Vegas Boulevard of all streets, out hunting for a different Gunnar while talking to his daughter on the phone. I really missed Brusewitz – he would have been a great help in the Widforss case.

'I'll get in touch when I get home,' I said.

'Good luck,' she said.

. . .

The man on Saint Martin, who turned out to be the same sort of age as me, lived in a town in the south of Sweden. I went down there as soon as he arrived home. It all seemed a bit too good to be true, but he was just as obliging as I'd hoped and he let me borrow his treasure trove of material without further ado. The first thing he placed on the table in the restaurant where we had arranged to meet was a virtually perfect copy of the poems *Songs of Yosemite* by Harold Symmes (1878–1910), published in 1923 and containing eight full-page colour reproductions of Gunnar's watercolours. This is much sought after by bibliophiles, and, although I knew it existed, I had never expected to see a copy. Nor had I expected to see a whole bundle of menus, once again illustrated in colour, from the Ahwahnee Hotel in Yosemite. And all the reviews of his exhibitions! It would have caused me endless trouble if I'd had to sift them out from the American press archives myself, but here they were, all gathered together.

The letters, however, were something of a disappointment. The originals had disappeared and what remained, rather strangely, were copies, or, more accurately, summaries: there were more or less detailed résumés of a couple of hundred letters and postcards from April 1910 to June 1934. A typewritten bundle of fewer than fifty sides compiled by Häggart himself in 1939, the year before he died, as if he wanted to preserve his friend's words for posterity – but not all of them. I got the impression that Carl Erik Häggart wanted to protect Gunnar. But from what?

That impression grew stronger when I began to read the typescript on the train on the way home. There was actually considerable similarity between Gunnar's letters to his mother and what he had written to his friend, but the gulf between hope and despair was much greater in this case. When everything was going his way and his joy in painting was shining through pure and clear, the light was that much brighter in these letters, as was his belief in his own ability; equally, however, when despondency had the upper hand the words here were deeper and darker. His business affairs sound consistently better in his letters to Häggart, but his loneliness is more profound. As I said, however, these sheets were copies, and there is no doubt that they had been edited. In some places there was the blunt statement: 'Private Matters'. And, of course, there was nothing I could do but accept it.

But the fact that gaps were being filled in was good enough in itself. It emerged, for instance, that Gunnar went to Skagen at the start of June 1920 and remained in Denmark well into the autumn, when he had an exhibition in Copenhagen. He had, moreover, acquired a guru.

My suspicions were confirmed: Gunnar's travels in the Tunisian desert had left him utterly exhausted. During the late winter and spring before his summer in Denmark he had been sending fairly neutral letters home to his mother in Stockholm, whereas his friend in Södertälje (which is where Astra was located) was receiving a version that in many respects was different.

'Stuck in Como. Taking a cure for neurasthenia.'

He seems to have worked himself up to some sort of nervous collapse. Run into a wall, as we say now. Hardly surprising, given his work-rate. And perhaps we shouldn't be too surprised when, willingly and amazingly successfully, he lets himself be treated by Emile Coué (1857–1926), the high priest of pop psychology. In a letter from Como in March 1920 Gunnar positively bubbles with enthusiasm when writing about this Frenchman, the man who came up with the concept of positive thinking based on auto-suggestion.

The method involved convincing yourself that everything was simply becoming better and better all the time. Coué, initially a pharmacist, had come to the conclusion that, while pills were no doubt good in many situations, there was nothing that could match imagination. If you believed in a medicine you immediately felt better. Ancient wisdom, and pretty simple at that. What was new, however, was to posit that the seat of the imagination was located in the human subconscious, whereas the will – the conscious intellect – was, in Coué's view, no more than a rather insignificant feature. He considered the discovery of the subconscious to be fully comparable with the discovery of America. It changed everything. Forever. His whole method thus rested on two principles so elementary that any fool could understand them. First, if there is conflict between the will and the imagination, the latter will win, without exception. Second, the human imagination is easier to control than the will.

In the 1920s, when Coué's fame was at its height, he had long since ceased to prescribe anything but autosuggestion. And many people saw the light, both when physically present at public seances or as a result of reading his book *Self Mastery through Conscious Autosuggestion*. It was a bestseller. That, presumably, is where Gunnar learned to chant Coué's famous mantra: 'Tous les jours à tous points de vue, je vais de mieux en mieux' ('Day by day, in all ways, I become better and better'). The Swedish version made its first appearance in 1923.

Like most successful teachings that offer salvation, Coué's method was a well-judged mix of the fraudulent and the self-evident, presented to some extent in the form of metaphor. The more banal the better as, for instance, in the case of the long plank. My absolute favourite.

'Let us assume that we lay a plank on the ground.

It's about ten metres long and thirty centimetres wide. It is obvious that every one of us is capable of walking from one end of the plank to the other without stepping off the side. But now let us change the conditions of the experiment and imagine that this plank is placed high up in a church tower. Which of us is then capable of walking even a small way along this narrow path?' We are not told how we are supposed to get this enormous plank up into the church tower in the first place, but the issue is quite clear: all the willpower in the world would not get me to walk along it, the very thought of it makes me dizzy. Only once have I been up a church tower and I shan't be going again.

With this trivial notion as his starting point, it wasn't really very difficult for Coué to make his thesis that our capabilities are governed by our imagination sound plausible. What stops us falling off the plank at ground level is our conviction that the task is easy. Not belief, but conviction, knowledge. We are convinced that we can and therefore we can. According to Coué, virtually everything we want to achieve is like walking the plank. All that is needed is autosuggestion. And should you have religious inclinations it's easy to slot the name of God into the mantra that states that everything gets better and better.

If I ever get round to organizing our fairly shambolic garden, I shall locate a plank ten metres long and thirty centimetres wide and place it as a walkway between the beds. Then I'll walk back and forth across it every day and think of Coué and of Gunnar's new-found courage and certainty about his own strength. And maybe remember, too, the musical revue artist Ernst Rolf, who stole the idea for his jingle 'Better and Better Day by Day'.

Perhaps it was the lunacy of the First World War that affected the 1920s and opened the road to patent panaceas of this kind. I don't know why, but, in all its simplicity, there is something desperate about the whole business. That anything that gets better and better must also inevitably come to an end was not a point Emile ever mentioned during his seances. In the case of Ernst Rolf, for example, all of his successes came to an inglorious end: when wife number three decided to move out, he tried to drown himself and then died of pneumonia at the age of forty-one. But that wasn't until 1932 and it's hardly relevant in the present context anyway. It's Gunnar we are talking about and now, aged forty-two, he moved to the United States.

Chapter 9

To the Land of the Chewing-gum Magnates

It has proved impossible to discover why Gunnar Widforss wanted to go to Japan, but that is what he consistently told American journalists throughout the 1920s when they asked him why he had settled in the United States. He was just passing through, he said, but had fallen so hopelessly and helplessly in love with the scenery that he had stayed. His goal at the start had been the Far East, Japan in particular, but his money had only stretched as far as California, where he found plentiful opportunities to paint and earn enough for a ticket across the Pacific. There was an abundance of scenes to paint, and also of customers.

In spite of that, or perhaps because of that, his dream of travelling around the world soon faded. Not even Hawaii was attractive any longer.

The art historian Osvald Sirén, who was the same age as Widforss, may well have been the man behind the interest Gunnar claimed to have in Japanese and Chinese watercolour painting. But it's just a guess. Sirén's book *The Golden Pavilion*, which dealt with his own years of study in Japan, appeared in 1919, and the winter before that he had been the driving force behind an exhibition of older Chinese painting at Stockholm University. In short, Asian art was in vogue and Gunnar, as always, had itchy feet. There may, of course, have been hundreds of other reasons why he set off. Chance. Flight.

What can we actually know about any human being's way in the world?

. . .

Gunnar arrived in California on 11 January 1921, having been travelling for four weeks. One of the first places he went to was the island of Santa Catalina, twenty-five nautical miles out into the Pacific Ocean southwest of Los Angeles. He was there before the end of the month. No one knows why Gunnar went there but it seems reasonable to assume that he had heard that the island was incomparably beautiful, and the most beautiful views were ultimately professional necessities to him. Writing to Carl Erik from the train up near Chicago just a couple of weeks earlier, he said: 'For almost two days and nights now I haven't seen anything beautiful.' The comment is typical of him. There is little that puts him in such a bad mood as feeling let down by the scenery.

He stayed on Santa Catalina for the greater part of February, painting a total of ten paintings that were then exhibited for sale in a shop on the island. His letters tell us that the results were meagre. After a couple of months in the country, during which time he had already discovered the Yosemite National Park that was to be his fixed point over the following years, he had managed to sell only one painting and that was a scene from Denmark he had brought from home with him. But let's not get ahead of ourselves. One of his first letters from America contains a comment that is impossible to ignore and that fires the imagination: it seems that the whole of Santa Catalina was owned by Mr Wrigley, the chewing-gum magnate.

I know a bit about islands and a little about money. Santa Catalina is some seventy-five square miles in area, bigger than Fårö in the Baltic and its location is good in that it's close to a city that even then was a metropolis whose streets were paved with gold. The island could not have been cheap, and I would guess that this fact provided our poor artist with a good introduction to the atmosphere and spirit of what was to be his new home. There are few things that tell us more about the United States in the 1920s than the chewing-gum industry.

Its history began in New York as early as 1869, when the Mexican general Antonio López de Santa Anna – also called the Napoleon of the West – happened to meet the inventor and entrepreneur Thomas Adams. General Santa Anna, who was seventy-five years old by this point, had achieved undying fame at the Battle of the Alamo back in 1836, when, as self-appointed dictator of Mexico, he had caused his army to crush the Texan separatists, including Davy Crockett, Jim Bowie and other legendary greats, who were killed to the last man.

Well, his victory was short-lived. The commander of the separatists, Sam Houston, who would have a city named after him in the fullness of time, struck back and Santa Anna was forced to witness Texas first breaking away from Mexico and then, in 1845, becoming one of the states of the United States. None of which is relevant here except as a backdrop to the observation that Santa Anna was a political survivor who won and lost power in Mexico repeatedly – sometimes with the support of the United States – and was eventually given political asylum by his former enemies, the Americans.

At the time he came into contact with Thomas Adams in New York, he was in the process of putting together the finances for yet another of his doomed Mexican revolutionary armies by selling, among other things, a large quantity of chicle, a type of gum harvested from tropical trees of the variety Linnaeus named *Manilkara zapota*, sometimes known these days as the sapodilla plum. Adams was, as we've said, an inventor and he bought tons of the stuff with the idea of processing it and selling it as raw rubber that could be used to produce tyres for wagon wheels, the point being that rubber was expensive whereas chicle was cheap. It was a brilliant idea, apart from the fact that chicle was not suitable for tyres, not in any way at all, which was a disappointment for Adams; but nevertheless this proved to be his fortune.

He discovered instead that chicle was the perfect base material for chewing-gum, and within a very short time the market for the older types of chewing-gum that Americans had long been addicted to, that were based on paraffin and the resin of pine trees, was dead as a dodo. The industry grew quickly and when William Wrigley Jr came on the scene in the 1890s the business was already enormous and there were any number of chewing-gum manufacturers competing with one another for customers.

Wrigley's father was a Chicago soap-hawker who realized that his sales increased when he gave his customers a packet of baking-powder as a gift when they bought his soap. The baking-powder proved to be so popular that he went over to selling it instead and now he gave customers a gift of chewing-gum. Then, in turn, he discovered that chewing-gum was a bigger hit than baking-powder. The rest is history. His son William built his empire on the basis of the slogan: 'Anyone can manufacture chewing-gum, the problem is selling it.' William was good at both, especially the latter. His marketing was both aggressive and innovative. Huge billboards at the roadside and free samples for millions of children were just the start. Later on, Wrigley invented the musical advertising jingle for the radio, and the annual bill for his neon signs in Times Square in Manhattan was $108,000 throughout the 1920s. On his death at the age of seventy in 1932 he was one of the ten richest people in the United States.

There were, of course, a number of lucky circumstances that contributed to his success, particularly the First World War and, before that, Fletcherism, an immensely popular method of shedding weight named after Horace Fletcher (1849–1919), the overweight businessman who launched the idea. Losing weight was only part of it; Fletcher's miracle cure, which basically involved eating extremely slowly and chewing everything for at least half a minute, was thought to prevent most illnesses, to cure pain and nervousness, to promote thoughtfulness and to put a stop to meaningless chat at mealtimes. We can imagine that the last, at least, worked pretty effectively; carrying on a conversation with someone who is chewing the cud must be a thankless activity.

So chewing was in fashion, and during the First World War, when conditions on the Western Front were literally hellish, the American chewing-gum industry managed to convince the Red Cross to ship millions of packets across as emergency aid to the French. As for their own American troops, chewing-gum was already part of the daily ration, since it was considered to be a pleasant alternative to cleaning your teeth, was thought to quench thirst and to have a calming effect when under enemy fire. Given all this, it is hardly surprising that William Wrigley could afford to buy the whole of Santa Catalina, as he did in 1919.

Wrigley was also keen on birds. During the 1920s he built a bird park on the island, probably the biggest bird park in the world. Enormous aviaries with thousands of birds from every corner of the earth. But Gunnar had moved on to better hunting grounds by that stage, in the mountains farther north.

. . .

I have always suffered from vertigo. I can't even climb a pine tree these days. And when my curiosity leads me to be close to a steep drop I have to keep turning away or bending down to see whether there is anything at my feet, anything small and ordinary that is not life threatening to concentrate on.

It is rarely difficult to fill in all the edges of a jigsaw puzzle. But then you find yourself sitting there turning over coloured pieces that won't fit anywhere.

. . .

Gunnar's first letters from Yosemite, the national park in the Sierra Nevada a day's journey east of San Francisco, date from early March 1921. There are so many subjects to paint and they are so spectacular that right from the start Gunnar intended to stay until the autumn. There was money to be earned there. He needed several thousand dollars to continue his journey to Hawaii and then on to Japan. He bought a bicycle, lived in a tent and painted as never before. Later he moved into the Sentinel Hotel, in the heart of the Yosemite Valley, where he paid for his keep with five watercolours a month, which seems to me to be expensive, given that he could have sold them himself for $50 apiece. He wrote that a good riding horse cost $150. Just as a comparison.

There were plenty of horses around and all of them were scared stiff by Gunnar's bicycle. All the other animals were too. Cars, on the other hand, left them unmoved. And cars were mentioned more and more frequently as time passed: living in the United States without a car soon proved to be impossible.

Chapter 10

The Travails of the Camel Corps

Right back at the beginning I mentioned that at a critical stage of the expedition – outside La Posada Hotel in Winslow – the memory of singing myself hoarse in the top of a pine tree one summer's night long ago came to me out of the blue. It was an attack of homesickness, I think, or perhaps an attempt to overcome the recurrent sense of imbalance that afflicts me when travelling through landscapes in which I don't feel at home. Places where I can't even put a name to the trees. That probably explains it: there were a number of big deciduous trees outside the hotel and by then I could recognize the turkey buzzards perched in them, but I knew nothing about the trees.

The evening before a friendly couple in the restaurant of the Weatherford Hotel in Flagstaff had tipped us off that La Posada was not a place to be missed. Built in 1929, it was one of the Santa Fe Company's classic railroad hotels, and, since we were going in that direction on our way to the Petrified Forest and Monument Valley, we turned off the highway and followed Route 66, which leads to Winslow. Or through it, rather, as there are not many people who stop there. Nor are there many reasons to do so apart from the hotel, which is always fully occupied, and a souvenir shop that bases its whole business rationale on a single verse in an Eagles' song that every American knows by heart. The verse takes place on the very street corner occupied by the shop.

Well, we thought, when in America do as the Americans do, so when the suite they called the Frank Sinatra Suite in La Posada Hotel just happened to be free it really made our day. Johanna's day, anyway, and she immediately disappeared off into town to see things, experience as much as possible, talk to people, find out everything she could and have a good time – all the things I envy her for but can't actually handle myself. When she returned a couple of hours later she had even discovered the exact details of what it would cost us to buy half a block that had recently burned down – a great big hole in other words. It was the perfect place, she said, to develop something new, something – I gathered in haste – along the lines of an Italian piazza with shops, cafés, the lot.

But I'd already been off doing some exploration of my own. I'd been out for a walk, no more than round the hotel, but that was enough. I'd come across the longest goods train I'd ever seen, that was for a start, and then I'd also seen a historical monument, if that's what you could call it: an ugly stone plinth with a memorial tablet that briefly stated that this actual spot was a 'historic site' because in 1857 Lieutenant Edward F. Beale and his camels, a whole caravan of them, had passed this way. This was at a time when the camel was thought to have a future as a beast of burden in the American Army. As a military experiment the

Camel Corps struck me as having exactly the right degree of optimism to offer a key to understanding the warm and carefree attitude the Americans have to vehicles of all kinds. It wasn't solely about horses and cars. And I was already familiar with Edward Beale, since he too had a mountaintop in the Grand Canyon named after him.

All this started in 1855, when the camel lobby, with Jefferson Davis at its head, finally succeeded in prising out of the government the $30,000 of funding they needed to initiate the programme. The arguments were convincing. First, the United States had used a large number of camels very successfully in its short war with Tripoli – present-day Libya – in 1805; second, a few years before that, Napoleon's dromedary regiment in Egypt had demonstrated what the animals were capable of. The rest was simple textbook stuff: the terrain and climate in the southwestern United States were not dissimilar to conditions in North Africa, and it was reasonable to assume that the tools Napoleon had used when persecuting the nomadic Arabs would help in the wars against hostile Indian tribes. The decision was made and all that was needed now was to acquire camels.

The American fleet adapted a vessel, set sail for the Mediterranean and put into Tunis, where all the witnesses are agreed that they made two mistakes, the first of which was to announce the moment they stepped ashore that they were Americans with plenty of money and they wanted to buy camels. Everyone, absolutely everyone in the whole city, immediately had masses of camels for sale, all with minor variations on the theme 'best camel in Africa, very cheap'. The second mistake the military made was to buy the first available camels at whatever price the seller requested. The news spread like wildfire, after which there was no hope of anything like sensible business dealings. They were forced to leave Tunis with precisely three camels in the hold, two in such poor condition that they sold them off cheap to a butcher in Constantinople.

But, say what you like about Americans, they don't give up easily. They kept at it, in Turkey this time, and their bad luck continued, mainly because the cunning British had requisitioned no fewer than 8,000 camels for their war against the Russians in the Crimea. The supply of camels and the price of the beasts were consequently unfavourable, and the Americans couldn't find anything worth buying. After three months of fruitless cruising around the eastern Mediterranean, the vessel anchored in Alexandria, still with only one camel on board. When they sailed north again towards Smyrna, after six months in the region, they had still not managed to buy more than seven. The camel trade was turning out to be a very difficult business.

. . .

That the acquisition of camels demands time and patience was something even King Erik XIV of Sweden was aware of. His coronation in the Year of Our Lord 1561 did not really take off and become the perfect Renaissance shindig with the required Continental éclat for the simple reason that they had been unable to get hold of camels. No aurochs either, for that matter. They had to make do with bears being baited by English dogs and they couldn't even come up with a lion, though they tried.

It was not until a couple of years later, at the time of the king's much debated marriage to his servant-girl mistress, Karin Månsdotter, that a living lion was installed on the square outside the palace in Stockholm. Nearly a century later, during Queen Kristina's years on the throne in the 1640s, they managed to steal a lion in Prague, where the beast had belonged to Rudolf II, collector and emperor. Around about the same time the Duke of Courland made the queen the gift of a leopard, which lived up to all expectations to the extent that the treasury was soon forced to defray the funeral expenses of a maidservant the leopard had taken a fancy to.

Exactly what use camels were put to in Sweden, however, is not reported in any degree of detail in the annals. But it's quite clear that camels came and camels went. In the 1580s, for instance, there is a fellow by the name of Bugdan Tatter in the capital and his occupation is given as dromedary keeper. It seems likely that they were used in parades, and we know for sure that camels paraded through the streets when Karl X Gustaf came to the throne in 1654. There were also magnificent carriages equipped with jingling bells and mounted blackamoors – the latter no doubt also imported, probably forcibly. A couple of years before that we hear of silver being paid to a Turk in the camel trade in Stockholm and to a Pole who looked after the animals in the court stables. Elsewhere in the archives we are informed that Count Nils Bjelke successfully defeated a whole caravan of camels in a battle somewhere in Hungary – it's said to have been in 1687. This also involved an unfortunate Turk called Schabbasch Bjelke, whom Bjelke brought back to Sweden and, so to speak, converted to Christianity.

Schabbasch has had a better fate than might have been expected, since generations of myopic art experts have recognized him in the famous painting *Caravan with a Turk Leading a Dromedary* by David Klöcker Ehrenstrahl, which has been hanging in one of the rooms in Drottningholm Palace for the last 300 years.

. . .

The Americans eventually found what they were looking for. The answer, as always when it comes to military transactions, was a middleman. In this case his name was Hadji Ali, a young camel-driver from Syria whom the Americans were fortunate enough to engage in Turkey, after which they were able to return to America in no time at all with a shipload of (I imagine) startled camels, both dromedaries and Bactrians. Hadji Ali, who came to be known as Hi Jolly as a result of linguistic confusion, accompanied a second shipment the following year (1857) as adviser on camel matters.

The base was set up in San Antonio, Texas, and that is where Edward Beale entered the project as leader of an expedition with the task of investigating whether there was a viable route to California along the 35th parallel. That's how the caravan came to follow the route that later became Route 66 and to pass the spot that became the rather unprepossessing city of Winslow. On the whole the project was successful, for camels were capable of carrying very much more than mules and they tolerated all the hardships of desert and mountain with admirable equanimity. There were, however, certain disadvantages, such as the fact that horses shied and reared up when meeting camels, but, in spite of that, a proposal to buy a thousand more soon landed on the desk of the minister of war.

It was rejected. The Civil War intervened, the military had other things to think about, and Jefferson Davis himself became president of the breakaway Confederacy. That was the end of that, and there was no one who wanted to take on a project that had been initiated by a traitor. The railways were beginning to make camel-trains superfluous anyway. Some of the camels were sold; others were turned loose and disappeared out into the wilderness. Some ended up in circuses, some as beasts of burden in the goldfields, but their fate was already sealed and their reputation hit rock bottom, particularly after one of them frightened the life out of a mule laden with whiskey for the gold-miners of Hells Gate up in British Columbia.

In the end they were being shot as vermin, which was not so easy, given that the mountains in the western United States go on forever and camels are long-lived beasts that can get by on next to nothing. Many of them survived for decades. One, called Topsy, which had been around from the start, died in Los Angeles

Zoo in 1934, and others of a later generation were still out in the wilderness. The last wild camel was caught in 1946, but even ten years later reliable reports were coming in of the occasional animal seen down in the Sonoran Desert.

Hi Jolly also lived a long and varied life in the United States. A legendary pioneer. Beside the road from Phoenix to Palm Springs, the Arizona Highways Department has raised a large monument over his grave. A pyramid crowned with a camel. Americans like monuments. Topsy's ashes are buried there, too.

. . .

When the first Europeans arrived, the Indians were living in a variety of different cultures spread across the whole of the American continent. What happened next is well documented. A tragic story, so dark and barbaric that the idea of packing these unfortunate people into protected reservations as an alternative to slaughtering them actually seemed good, almost humane, at least when viewed as a makeshift solution. They began to mark out Indian reservations at the end of the 1860s, just a few years before the first national park, Yellowstone, was created in northwestern Wyoming. The same idea, the same mistake.

There are about 300 more-or-less autonomous reservations, and we drove right across the biggest of them. A whole day through a landscape that was flat and dry and somehow lacking in information, following dead-straight roads that seemed designed for sitting in the front seat of a car in a state of slow and silent contemplation. There surely can't be anyone who doesn't understand why the policy failed – the very idea, for a start, of converting heathen nomads into God-fearing agriculturalists, added to which the land they were apportioned was largely worthless. Nowhere else in the United States are the reservations more numerous and bigger than here in the southwest, and they are beautiful and poor, more or less like our national parks in Lapland.

It is not difficult to protect and preserve magnificent views. The only thing you need is money. Diversity, on the other hand, biological and cultural diversity, has a tendency to wither away when enclosed in enclaves. Apart from which, a bad conscience is scarcely the best guide.

Chapter 11

Painter of the National Parks

Every man's life is a labyrinth. Once you have found the way in, you can spend endless amounts of time there. Gunnar Mauritz Widforss may be quite special as an artist and unique in his choice of genre, but as a private individual he is hardly any more of a mystery than most of the rest of us, which in itself is enough of a riddle. A good deal of information disappeared forever when he died; much else has been dispersed and lost in the seventy years that have passed since his death. All those letters rescued from the skip merely served as confirmation that there would soon be no traces left. It had been that close. I could quite easily have passed him by. Just chance.

But perhaps it wasn't too late. Two questions – no, many questions – followed me on my travels, but two were more significant than the others.

What was it that drove him?

What right did I have to be rooting around in his business?

The second of those questions was never properly answered; nor, for that matter, was the first, although I thought I had some indication of being on the right track. And that I was right to be doing what I was doing, that poking my nose into it all was okay. It came in a letter he wrote in the first years of the Great Depression. The letter had been lying in an attic in San Francisco ever since, a long, desperately heart-rending letter about deceit and broken friendship. His best friend! The painter Albert Thomas DeRome.

Gunnar wrote to DeRome that he'd not had more than six or seven close friends in his life. They had come and they had gone, life is like that, and a couple of them were still there back in Sweden, at a distance, far too far away. But, most important of all, the heart of what he now felt he had to say, was that as long as he could remember he had needed a friend, just one friend, closer than all the rest, closer than all the other more or less superficial friends, colleagues and acquaintances. And who, I thought, who is that friend now if it isn't me? Which is why I entered the labyrinth.

Presumptuous? Yes, possibly. I didn't think about it so much at the time. Curiosity dulled my doubts. The time would come when I would have reason to regret my decision, and I would experience the dull ache of conscience that goes with being a grave-robber.

I had suspected all along that there was something waiting to be discovered. In addition to which I thought I could be absolutely certain that people don't tell their mothers everything. We had a long conversation about that, my mother and I, when I asked her to let me borrow the letters I had written to her during my own travels in the early 1980s, a thick bundle from eighteen countries, more often than not a jolly and straightforward account of the bare facts about the where and the when but seldom the why. No mention of fear or loneliness or the misery I was ashamed of; I simply reported what I had believed in advance, or what I had hoped to experience, rather than the disappointments that had actually happened. As if I were sending home photographs of beautiful landscapes.

. . .

Albert Thomas DeRome (1885–1959) was in advertising and he loved fast cars. He was the head of marketing at a chocolate factory in San Francisco, earned money, had a family and continued to cherish his youthful dream of being a painter, a proper artist. In that respect he may be seen as a worthy predecessor of all those present-day copywriters, art directors and more-or-less gimmick-ridden advertising people who, deep down inside, want to devote their time to something more important than serving as well-paid infantrymen in the assault mounted by commercialism on everything that is serious and thoughtful. It is hardly surprising that he and Gunnar found one another. Albert gained a teacher and Gunnar an agent, a fixer with contacts and an understanding of commerce and worldly matters. And, moreover, I'm pretty sure they had fun together, like a couple of boys on the run in the light nights of summer.

As far as we can judge, their friendship was already well established just a few months into Gunnar's first year in California. They probably met in San Francisco, when Gunnar was scouting for art dealers, or perhaps it was in Pacific Grove, south of the city and outside Monterey and Carmel, since that was something of an artists' colony at the time and one of the places Gunnar would frequently return to over the years. At all events, it is clear from the letters that they set off in September for a month's road trip together up around Mount Shasta, close to the Oregon border. Gunnar writes of beautiful mountains and half-dead settlements with names like Peanut and Hayfork in the old gold-mining district, but he also writes that the light in the Sacramento Valley is so wonderful that 'the most insignificant scene became as beautiful as a dream.'

DeRome owned a roadster, a two-seater sports car, which – according to Gunnar – he drove with great skill. 'The fastest we drove was 45 miles an hour, but I had to act as lookout and keep watch behind us in case any "speedcop" was approaching.' The fellow seems to have driven like a car thief. And they did indeed crash on a corner, which led to smashed cars, hard words and lawsuits all round, but since DeRome was a keen photographer who carefully documented his crashes he had no difficulty in winning his case. The car was patched up and they continued cheerfully on their way. I have seen some of his photographs, including one of this particular crash and another of Gunnar standing beside a half-tame grizzly bear. I get the impression that they were enjoying happy days, truly happy days.

Why it went so wrong is hard to say. Perhaps they fell out about money, about loans that weren't paid back on time, or perhaps with the best of intentions Gunnar was a little too harsh in his criticism of DeRome's hit-or-miss painting. The letters offer hints of both. For whatever reason, and to Gunnar's genuine distress, they went their separate ways in 1930. One of DeRome's relatives, a woman in San Diego with whom

I still remain in friendly contact, has told me that the cause of the break-up may have been that Albert's wife, Martha, simply couldn't stand Gunnar. There is a story in the family that on one occasion, when he had been invited to dinner with the DeRomes, he first of all drove Martha mad by chain-smoking, a bad habit he'd had since his youth, and then thanked her for the meal by saying in passing 'that he really didn't like family life'.

But they were close friends throughout the 1920s, and I think the pleasures and the benefits were mutual. There is no doubt that Gunnar's successes at the galleries in San Francisco, Oakland and Los Angeles can partly be ascribed to DeRome, who, in turn, eventually became a real painter. Not like Gunnar, but good enough to make a name for himself. Typically, it would take another car crash to get him to say goodbye to the chocolate factory. It was in 1931. His last crash. He broke his neck but survived, dumb and partially paralysed. The only bit of luck was that his brush hand was intact, and from that day on he was an artist. He could easily have been as forgotten in his country, as

Gunnar was in his, but the insurance company had told him he could do what he liked as long as he didn't earn money. So he never sold his paintings. They were all stored in his studio and inherited by members of the family. It was only later that the art lovers of California discovered him. Exhibitions. Monographs. Fame.

There is also a monograph on Gunnar, of course, published by the Museum of Northern Arizona in Flagstaff as late as 1969 in connection with a retrospective exhibition. *Gunnar Widforss: Painter of the Grand Canyon* is a beautiful large-format book with expensive colour plates and appreciative text. It is now impossible to get hold of it. When it does turn up in some remote second-hand bookshop, it is priced at thousands of kronor. It's not that much more expensive to go to Flagstaff and see the paintings for yourself. They have several of them and they are actually on permanent exhibition. A fine museum, in a nice town about the size of Kalmar.

But Johanna and I never went to Yosemite. The only thing I wanted to do in the United States was to hike the Widforss Trail and see the Grand Canyon with my own eyes. Some birds, too, perhaps, but that's all. It was Yosemite that was Gunnar's university, though. A landscape hard to paint, with which he struggled during his first years and eventually taught himself to see and understand as no other artist did. And for once he was in the right place at the right time.

Tourism was big business in the 1920s. In California it was actually the second biggest business, and both tourists and those exploiting them were only too keen to buy art. It's said that the building of the Panama Canal, which opened in 1914, had a major impact on the railroad companies and forced them to do whatever they could to replace the lost goods traffic with tourists. It was no accident that Sven Hedin visited the Grand Canyon at the invitation of the Santa Fe Company, which also gave Gunnar many commissions, as did the Union Pacific and other companies, often in close cooperation with the hotel chains, which were throwing up imposing imitation baronial piles in the national parks. The Americans were beginning to discover their wilderness in earnest, and the California

Watercolor Society had just organized its first exhibition in Los Angeles.

Gunnar was old enough and wise enough to see his chance when it arose. He knew that artists needed to specialize if they were to make their mark, and here there was a market for what he knew better than anyone and wanted to do anyway. As far as I can see, there was seldom any need for him to examine his conscience. He touches on that sometimes in his letters, and, even though what he wrote about it now only exists in a rhapsodic summary (or perhaps *because* it only survives in that form), it is worth quoting. Here is an extract from the end of a long letter written in March 1923:

> *As far as possible I'm trying not to paint views just to sell them. Paint 7 or 8 canvases a year for business. In between, periods of rest in San Francisco, Santa Barbara or Los Angeles. Last winter 5 paintings a month for 5 months. Reckon on around 12 paintings a month during the summer months. At the moment Mr Mather in Yosemite is the boss of America's national parks.*

Interested in my paintings. Bought one two years ago. He wants me to go to Bryce Canyon, which he wants to incorporate into Zion National Park. Wants paintings of it for propaganda purposes. Have applied for membership of the Sierra Club.

Afraid of getting fat (if I use a car). Private matters. Go on a diet sometimes. Blood pressure too high. Smoke c. 18 cigarettes a day. Get a drink or, even better, two from some kind soul. Write about prohibition. People think that beer and light wines should be exempted. Learned to dance a year ago. Am just as good a dancer as anyone else. And if the girl is good, it's a real pleasure. And good exercise.

That last bit is typical of the time and reminds me of Ivar Broman, the physiologist and Darwinist. In the same year as Gunnar's letter, Broman wrote an essay in the Gothenburg newspaper *Göteborgs Handelstidning* with the title 'The Importance of Dance as Physical Exercise': it gives a good picture of the ideal of health that the people of the day were seeking in virtually every aspect of existence.

Broman had happened to come across a report published in the proceedings of the Finnish Medical Association that dealt with the comparative energy expenditure of various kinds of dance. The waltz and the shimmy, for instance, were found to have roughly the same energy consumption as a washerwoman working hard, and that in turn was about double the amount used on an ordinary walk.

Stimulated by this surprising discovery, the Finns had cheerfully continued their research and after thorough experimentation concluded that the foxtrot and the schottische approached the effort expended by a stonemason, while 'anyone dancing the polka or the mazurka is working harder than someone sawing wood.' Strangely enough, the tango does not figure in the material. I suppose the Finns were more addicted to the mazurka in those days.

Ivar Broman was a man of wide interests and sufficiently open-minded to tackle more mysterious questions than dance. His investigation 'On the Utility of Snoring' is particularly memorable. Unfortunately, it doesn't really have a part to play in the present context and I don't suppose dance does either, but when you're doing a jigsaw all the pieces count. Mr Mather, for instance. There was no one more important when it came to Gunnar's career.

Stephen Mather (1867–1930) put the world of business behind him and left it with millions in his pocket. Then, as Americans do, he turned in a more idealistic direction and focused the energy for which he was already legendary on the development and expansion of the national parks. When the US National Park Service was set up in 1916, he became its top man, and by the time he relinquished that post shortly before his death the area under the protection of the National Park Service Organic Act had more or less doubled. This was the golden age of American national parks. A number of them had existed for many years – Yellowstone ever since 1872 – but they did not have many visitors. Accessibility was often dreadful. Bad roads and bad hiking trails, no hotels and hardly any campsites.

There were, of course, many factors that contributed to success. The ground had been well prepared in advance by prophets such as Thoreau and John Muir, but there was also the fact that Stephen Mather set to work at a favourable time: car ownership was becoming more and more common, and Americans wanted to see their country, especially the most beautiful parts of it. The preservation of nature was a political issue ripe for the picking, particularly by a man who knew how to go about it, and, as always, the first step was to demonstrate what was there to be seen and preserved. Beauty. And once Mather recognized the talent of the little Swede, he moved fast. Gunnar's first commissions outside Yosemite were Zion and Bryce Canyons in southern Utah.

Within no more than a couple of years and mainly thanks to Stephen Mather, Gunnar was able to establish himself as a sort of court painter to the national parks. No one had a better feel for their beauty. Initially he was mostly based in Yosemite, but later the Grand Canyon became the base from which he ranged in all directions. By early in 1923 he was selling well enough to afford a car, a second-hand Overland, and in the same year he was able to trade up to a Chevrolet

Roadster, still second-hand but a better car. He bought it from Ansel Hall, a legendary figure within the National Park Service and one of the people who gave Gunnar commissions to paint pretty well all the national parks in the western United States.

His travels took him to Mesa Verde in Colorado, Carlsbad Caverns in New Mexico, Crater Lake in Oregon, Sequoia and Death Valley in California, even up to Yellowstone in Wyoming. Not since the glory days of the great national painter Thomas Moran (1837–1926) at the turn of the century had there been watercolours of this quality, and there were people who considered Gunnar to be the best of the lot when it came to reading the nuances of nature and portraying the grandeur as it really was. Moran belonged to a different age, influenced by the romantic Düsseldorf School of painting, whereas Gunnar had followed his own path, relatively uninfluenced by schools and other more or less short-lived changes of taste.

There were people who sometimes criticized him for excessive naturalism and might have given him cause to feel hurt and dejected, but, according to most accounts, he was not particularly affected by criticism. His task was to reproduce what he saw but without being a camera. No one understood better than him how difficult it is to capture the soul of a landscape, even if it is just a pine in the sun of high summer. And he has lasted. The Grand Canyon provided him with his test-pieces. Anyone leafing through one of the magnificent volumes containing all the artists who have tried to portray the national parks will quickly recognize that it is Moran and Widforss who have best stood the test of time: the former because of his dramatic magnificence, the latter for his subtle treatment of colour and contours. Perspective, above all perspective. He saw features that no one else did, not painters anyway. Others, perhaps.

It may be significant that it was an old, yellowing, black-and-white photograph of one of Gunnar's watercolours of the Yosemite Valley that made me see the connection with Ansel Adams. The picture was in a bundle of photographs lent to me by the man on Saint Martin – large-format contact copies that Gunnar had sent home to his friend Häggart in Södertälje. The date of the photograph is not known, but the picture itself is dated 1924 and shows Half Dome, one of the most depicted mountains in America, in every medium. Some trees in the foreground, a rock, long shadows. I don't know where the watercolour is but the picture is all I need and more, for it places Widforss alongside Adams and prompts thoughts about art and landscape, and then art again.

Ansel Adams (1902–84) visited Yosemite for the first time in 1916, when he was in his teens, and it is claimed that the visit determined his career. That, of course, might well be an idealized later interpretation, but no one can deny the growing acclaim that met many of his photographs, of which the very first (from 1926) is called *Monolith. The Face of Half Dome, Yosemite National Park*. The same mountain, the same idea, a different medium.

We don't know whether Widforss and Adams knew each other, but they may well have met, especially in view of the fact that both of them worked for Ansel Hall – the chap with the Chevvy – who played a part in Adams giving up his career as a pianist and becoming a photographer instead, or so the story goes. At all events, he was the man who more than most saw to it that Ansel Adams took over the role of being the foremost artist of the national parks after Gunnar's death. And no artist has had more significance than Ansel Adams for the American appreciation of the wilderness as an environment worthy of love and protection. Adams's position as by far the world's greatest landscape photographer is likely to endure for a very long time.

That is the measure against which Gunnar's painting has to be measured. He and Adams were drawn by the same motifs and they worked with similar methods. Their genius lay in their eye for a motif: the rest was technique and waiting for the light. Naturalism or not was a meaningless question.

The main difference perhaps lies in the opportunity the photographer has to choose from hundreds or thousands of negatives, whereas the painter who needs to sell to make a living is under pressure to succeed every time, which is impossible. So he always deserves our forgiveness. Photographers also have the advantage of owning their pictures, which means they have access to them and can produce copies for exhibitions and books. The painter of watercolours, on the other hand, almost always parts with the original, particularly if he is not good at managing his finances. Gunnar refers to this disadvantage on several occasions: his best paintings were already sold by the time an exhibition was in the offing.

I have seen some of his paintings and they are as good as Ansel Adams's best photographs. Great art. It's not easy to locate them, but it can be done. Another journey.

Chapter 12

My First Inside Pocket

Gunnar's cooperation with Stephen Mather and Ansel Hall in the National Park Service brought him many advantages and they weren't only of the commercial variety. The use of art to promote the national parks also led to corresponding exposure of the artist himself. His watercolours appeared more and more frequently in magazines like the *Literary Digest* and the *World's Work*, sometimes as full-page reproductions, sometimes on the cover. In addition to which, previously unsuspected opportunities to exhibit in public spaces emerged.

Over the years there were many exhibitions, both large and small, mostly in the cities of the West Coast and in the new hotels in the national parks, but also in New York. The biggest of all was at the National Gallery of Art in Washington, D.C. in December 1924. Seventy-two watercolours were hung in all, most of them large-scale works and almost all new. Gunnar had been painting like a demon all year, and for once he had even been able to afford to hold on to the best pieces. The exhibition was consequently a great success, a breakthrough more valuable, economically and otherwise, than the Paris Salon twelve years earlier. The review in the *Washington Post* quoted prominent connoisseurs as saying that this, good people, is the best in this genre to have ever been painted in the West. The *Sunday Star* was also profuse in its praise, as was the correspondent of the *Los Angeles Times*. The event was even noticed and remarked on by the magazine *Vecko-Journalen* back home in Sweden. Gunnar was now forty-five years old, but he looks older in the photographs taken at the show's opening.

One of the works in the exhibition, a painting of Zion National Park, is still owned by the gallery; another is the panorama of the Grand Canyon that later ended up in the National Museum in Stockholm. The rest were sold, not just then, perhaps, but bit by bit, and so they disappeared. Any attempt to trace them with the help of the exhibition catalogue, comments in the correspondence and photographs from the gallery would be the labour of Sisyphus, but possibly worth the trouble. I have already found the painting listed in the catalogue as *Half Dome, Yosemite National Park*, although only at second-hand, as a photograph. But that's fine. I almost always find black-and-white photographs more pleasing than colour pictures. They don't give you too much information.

Nor, because of my own stupidity, does my American travel journal. There was something about the smell of it, I think. The first pages contain a small number of barely legible notes – difficult to work out in other ways, too. The rest is nothing but empty pages. I also took a camera with me, a smallish digital, but it didn't get used much either. A couple of photos of the car, a few trees against the light and a blurred butterfly.

It always ends up like that. I'm forever buying better

notebooks, which I then don't bother to make use of, though I keep them ready at hand. It's an old habit I've had ever since I got my first jacket with an inside pocket. It was a dream come true. I'd been longing for a jacket like that for ages. It didn't matter what it looked like and it didn't matter if the fit was as bad as a suit on a scarecrow as long as it had an inside pocket – that was the absolute tops as far as I was concerned. I never told anyone about this dream, which is probably why the whole business dragged on longer than necessary, but one fine day I inherited the real thing as a hand-me-down from an older cousin. I walked around for a whole day with my hand inside the jacket. I remember it being a little uncomfortable, but it was worth it because that was the moment at which I thought I understood something essential about adult life.

I soon began to wonder what I could keep in the pocket rather than my hand and that's when I came up with the idea of getting a notebook. That's how it started and that's how it still is. I must have been about ten.

Many years later I was lucky enough to get hold of a notebook with a smell that reminded me of the extinct quagga in the Museum of Natural History in Stockholm. The smell was identical, molecularly exact. The connection between stuffed animals and certain notebooks manufactured in Shanghai is presumably not as far-fetched as it might seem: the glue in the binding is probably produced from hoof, horn and bone, but what I do know is that in this case the smell was so exact that the book kept me company for a long time and I eventually filled it to the very last line. That rarely happens. Usually, as I said, I just fill a few pages and then buy a new one.

The quagga at the Museum of Natural History, one of very few specimens of this now extinct zebra of the veldt, had always captured my imagination. The other stuffed animals, too, for that matter, and the larger birds, but above all it was the quagga, which was actually quite a small foal. I sometimes wonder what there is inside it. The stuffing, that is. Wadding and straw, I suppose, though I don't have any real idea and don't actually want to know: I'm happy with the thought that explorers in the old days, when pushed, used whatever came to hand. Moss and newspaper. And why not other things as well? Unfinished love-letters and travel narratives that were going nowhere? The kind of things people wanted to forget but lacked the courage to burn. No doubt the bigger animals are hollow.

A year or so ago I had reason to spend several extended periods of time in the museum. It was in a department not open to the public, and every morning I walked past a full-grown giraffe that stood dusty and moth-eaten in an otherwise empty and deserted stairwell. We can assume that it was the height of the ceiling that had determined its location. Antique giraffes are difficult to house, and if they end up in a place where there are no interested visitors they can look very lost. Empty and pointless. Not worth bothering about.

What was special about the giraffe in the stairwell was that it had a black plastic bag over its head and appeared to be standing there thinking about some important issue. As if locked into an attempt to summon up memories.

The reason for my visits to the museum was that I had agreed to provide the skeleton plan for a major exhibition on climate. It was a hopeless task and I didn't get very far. They probably thought my proposals unworkable, stupid even, but I'd been given a free hand and I'd aired the notion of exhibiting reproductions of Brueghel the Elder's scenes of ice-skating on the Dutch canals during the little Ice Age of the sixteenth century – interesting from the point of view of climate history. Also, perhaps, one of William Turner's peculiar sunsets, together with other light phenomena he painted: connoisseurs of art ascribe them to his powerful visionary imagination, whereas more scientifically minded killjoys insist that the sky actually looked like that as a result of the major volcanic eruptions.

Climate history can be genuinely exciting and you quickly learn that there is nothing new under the sun, but my patrons wanted a rather more apocalyptic approach that focused on human stupidity and melting glaciers. I tried my best, wanting to be as accommodating as possible – I really did – but it was hard. It's partly that I'm not completely convinced that the climate changes we see around us can be explained as simply as is usually done; and it's partly because of the impossibility of setting up an exhibition of which the underlying tone is one of fear and horror.

I was saved by the giraffe. My idea was a simple one. Since stuffed animals per se can be expressive, I suggested that we ask the museum's famously skilful taxidermists to create a full-scale lifelike copy of a woolly rhinoceros, which could then be placed right in the middle of the enormous exhibition hall with a very small label as the only comment. No more than the name of the species, a short piece of ecological information and a brief sentence to the effect that it died out because of climatic changes 10,000 years ago. That was one thing. The other element – and the real point of it all – was to have a big, gleaming, irresistibly beautiful Cadillac in perfect condition standing alongside the rhinoceros. It would be a late-1950s cabriolet, preferably lime-green, the ultimate dream car. That, too, would be labelled with no information beyond its name and maybe a sentence about its fuel consumption.

People could work out for themselves, I thought, that the car would die out for more or less the same reason as the woolly rhinoceros. And with that I considered my job to be done.

Nothing came of it. The exhibition was mounted, of course, that had already been decided, and school classes attended compulsorily, I assume, but the curator omitted both the car and the rhinoceros and that was the end of that.

'Don't forget the woolly rhinoceros.'

A typical entry in my notebook. You can see from the shaky handwriting that it was written in the Mustang, probably on the way to Flagstaff, probably when we were level with Humphreys Peak, the long-extinct and snow-covered volcanic cone visible on the southern horizon from Widforss Point and many other places around the Grand Canyon. The forest was dead. Great swathes of mountainside were covered with dry dead trees. In their thousands. Whole forests. What had happened to them?

Later, when we were leaning on the bar in the Weatherford Hotel in Flagstaff, a local man told us that the whole region had been suffering from appalling drought conditions for several years, a climate change possibly as severe as that which forced the indigenous population to desert Mesa Verde in the thirteenth century. Drought stresses vegetation; insect pests complete the destruction. There are traces of fire everywhere.

It's quite clear that the undeniable climatic changes of our age have to be taken very seriously, whatever the cause might be. The Mesa Verde, that remarkable hill in southwestern Colorado that was designated a national park in 1906, provides a frighteningly good example. Gunnar was there in 1925 to paint the strange dwellings in the hillsides that had been excavated and carefully documented (some say plundered) by Gustaf Nordenskiöld (1868–1895), the forgotten son of the polar hero. Nordenskiöld had studied palaeontology at Uppsala and then contracted tuberculosis on Spitsbergen. He was on his way to Egypt for the change of air that might possibly help with the disease when, on a whim and for reasons unknown, he set off for Denver, Colorado, instead and while there picked up a rumour of a lost civilization on a mysterious mountain in the southwest. The rest is history. Gustaf Nordenskiöld's book *The Cliff Dwellers of the Mesa Verde* is a classic. It appeared in 1893 and he died, aged twenty-six, the following year. Strangely enough, no one has yet told his story.

Anyway, his memory is preserved at Mesa Vede with as much respect as Gunnar's is at the Grand Canyon. The Americans are good at things like that, they really are.

We were staying at Far View Lodge in the national park, a place that certainly lived up to its name. I'm sure you could see sixty miles from the balcony where I sat in the sunset with my elbows on the rail watching for birds at 8,200 feet above sea level. A hen harrier flew past, heading north, flying low across the mountainside. There were blue-grey gnatcatchers and ash-throated flycatchers. A restful end to an eventful day. And what happened later during the night was, of course, no more than a trifle. Among the myriad impressions of that day in the sun I can't offer any reason at all as to why such a bagatelle ended up being the only entry in my notebook.

It was the middle of the night, dark and silent. Not absolutely silent, however, as the hotel was not fully soundproofed. Other guests came and went. A car door was shut out in the parking lot. Someone was coughing in the room next door. Johanna was asleep and I was lying there thinking about nothing in particular. After a little while the coughing started again. I got up and put on my pants. It sounded like a smoker's cough, I thought, of the sort that only really obdurate smokers have. Chain-smokers, people who even smoke in bed at night, with everything that implies. The risk of fire, for instance. And who wants to run naked out of a burning hotel?

My mind always runs ahead of itself like that, making me cash in everything in advance, whether it be pleasure, misfortune or whatever. Experience has no say in it, however much it tells me that virtually none of these things ever come to pass.

Perhaps it was the smell that was haunting me, for the trees on Mesa Verde had only recently been destroyed by uncontrollable fires. They were virtually all gone. The simple truth is that I'm no good at travelling, that I don't have the right disposition to get any real benefit from it, but I still find it impossible to deny the value of seeing things on the spot. And smelling the smells. You can spend all the time in the world in your library thinking about the landscape and the future and coming up with theories, but there is nothing that beats seeing things for yourself without any intermediary.

In the long run climate changes may well lead to the disappearance of reserves and to their replacement by something better, something more flexible and more lasting. There is much more in need of preservation and protection these days than there was in Gunnar's day, when beauty was the prime reason for the creation of reserves, in Sweden as elsewhere. Mighty mountain peaks can take most things that are thrown at them – storms, heat, sudden changes of all kinds. Ecosystems, on the other hand, forests and wetlands, with all the diversity of life both great and small they contain, are merely imprisoned in reserves. Caught as if in a trap when the heat or the cold comes. When change comes.

There's nowhere for them to go. Sundays are treacherous. The age of reserves is over. I may be wrong. If so, that's the only thing I'm not frightened of.

Chapter 13

Why Natural Parks?

We stayed in pleasant little motels along the way, in Cortez, Colorado, and north of there in Moab, which is on the way to Salt Lake City. There is not really much to tell. The Americans were kind and pleasant and never anything but helpful. In a tiny settlement called Bluff we saw hummingbirds, which was new and worth noting. As was Bryce Canyon, though even that can't tempt me to try to say anything about it. We arrived at sunset and, as usual, everything was hot and indescribably beautiful. Well, indescribable anyway. A little unreal.

The road there that same evening, however, certainly

is describable, even though the scenery was the most beautiful of the whole trip. Ordinary farmland. We turned off the highway – Interstate 70 – and drove south along the valley through which the Sevier River alternately rushes or meanders in a leisurely way. A stroke of luck, perhaps, an early summer's evening with the sun breaking through after gentle rain. The June green of the riverside meadows was enchanting. And the scents! Scattered farms, tractors in the fields, cows grazing. A patchwork, a living farming landscape, the loveliest evening of the year: everything seemed to have sprung from an essay by Thoreau. We stopped at a lodge in a bend of the river, where a belted kingfisher was flying over the water and a night heron was sitting as motionless as a statuette on a branch in the bushes between the road and the riverbank. The coffee was weak – as always – but the sense of nature and culture in pleasant harmony with one another was all the stronger. Everything was a work in progress, at the halfway stage, so to speak; nothing was absolute. When all is said and done, is there anything as sad as wilderness without people working?

Don't misunderstand me. I admire the Americans for their desire and their ability to care for their national parks. We could see it for ourselves all the time: clean, tidy, careful. But the following morning, when we returned to Grand Canyon Lodge on the North Rim in order to follow the Kaibab Trail down to Roaring Springs and Ribbon Falls halfway down to the floor of the canyon, a crack in the façade suddenly opened up. I was the only one to see it, but it made me suddenly remember a dissertation from the past. And an unhappy person.

We were going to stay down at Roaring Springs for a couple of nights with an artist we know, but this was in the morning before we began hiking down. I went out to Bright Angel Point for a stroll. It wasn't far. The whole of the Grand Canyon lies at your feet and that was when I caught sight of some birds far down in the ravine to my left. Western tanager and Steller's jay, the former orange-yellow with a shrill song, the latter cobalt-blue and screechy, and I followed their movements in the tops of the tall trees three hundred feet below through my binoculars. The perfect harmony was broken, only a little but broken anyway, by a plastic bag that had been blown around and then got caught up in a Douglas fir a little way from the trail. This bag proved to contain two stories, one tragic and one rather more amusing.

The first has to do with a dissertation, an examination piece written in 1977 for the Agricultural University, as it was called then, in Alnarp and the title was 'Why Nature Parks?' A couple of hundred pages, simply stencilled and stapled and bound with black insulating tape. On the cover there was a long quotation from an American novel:

> *We arrive at the turnoff to Crater Lake and go up a neat road into the National Park ... clean, tidy and preserved. It really shouldn't be any other way, but this doesn't win any prizes for Quality either. It turns it into a museum. This is how it was before the white man came ... beautiful lava flows, and scrawny trees, and not a beer can anywhere ... but now that the white man is here, it looks fake. Maybe the National Park Service should set just one pile of beer cans in the middle of all*

that lava and then it would come to life. The absence of beer cans is distracting.

This passage is to be found on page 317 of Robert M. Pirsig's *Zen and the Art of Motorcycle Maintenance*, a cult novel published in 1974 that I've tried to read several times without either understanding it or managing to finish it. But you don't need to travel very far in the western United States before you realize that there are still many people who sleep with that particular book under their pillow. Motorcycles everywhere. Elderly and middle-aged men sporting leathers and ponytails on gleaming Harley-Davidson Baggers with biker-girls of a younger model on the back. They usually travel in flocks and, as a matter of principle, every one of them looks identical. Even the small bandanas these gentlemen tie around their heads so artistically are of precisely the same cut. It's important to belong. In their everyday lives the motorcyclists are presumably well-paid salaried employees who all wear the same tie. It's conceivable that they need to have a common text to discuss and interpret during the evening, and we can assume that virtual incomprehensibility is actually an advantage.

For my part I don't know much about philosophy or about motorcycles, but my grandfather did have a Harley-Davidson. Dad showed me a photograph taken in 1922 of him at the age of two sitting in a sidecar with a broad grin on his face while my grandfather posed proudly beside him. The latter was born in 1884 and died when I was small, so I don't know much about him apart from that he sold the Sjöberg brewery on Barnarpsgatan in Jönköping quite early on and thereafter painted amateurish but sincere oil paintings depicting horses. According to family tradition he was a very good-natured man, almost always happy and content.

Anyway, back to the stencilled pages from the Agricultural University. They had been written by a young Dane called Astrid Slottved. I had never heard of her but I came across a copy of her work by chance, at a point when I had just lost my childlike faith in ecology and ended up doubting everything I had previously considered sacred.

I won't go into it, not just now anyway, but I do want to give credit to this woman, the first in Sweden to ask the question why. I don't agree with all of her analyses, which are rather overladen with the far-left jargon of the time, but even then, fifteen years ago, I was both astonished and pleased by her view of the nature reserve as a mirror of a sick society. It was Astrid Slottved who made me see reserves not merely as 'Sundays' but as segregation. I immediately set about trying to find her. What had she gone on to do later? What did she think about reserves now? What was her present take on these scattered islands in an age dominated by monoculture?

The only thing I could find out was that she had taken her own life while still young. No one talks about her, no one remembers what she wrote. If only she had been a poet. Poets don't get forgotten so easily.

But one fine day her time will perhaps come and the politics of nature will be fuelled by something other than bad conscience. For what is it that actually distinguishes the shame we feel about the genocide that led to the creation of Indian reservations and other enclaves closer to home, what is it that distinguishes that from the shame we feel about the devastation wreaked on the beauty of forest and mountain during the industrial age? The contrition we feel about vulnerable plants and animals? In my world, the crisis of extinction – which is what currently drives the division of the landscape into Sunday parks and weekday sadness – actually points to something beyond nature reserves. The grey dipper we saw at Ribbon Falls, in the same part of the canyon in which Sven Hedin had unrolled his sleeping bag many years before, doesn't need reserves in order to survive, it just needs a little respect. That's true of most things, not only birds, but freshwater pearl mussels and everything else.

Amazingly enough, during our walk we also saw the California condor, which until recently bred only in captivity. It has turned the corner now. At the start of the 1980s there were only twenty-two left in the whole world and newspaper editors had the obituaries ready and waiting; now there are more than 250 of them. Americans again – say what you like, once they decide on something they usually manage to do it. A steady stream of condors was hatched in big aviaries, similar to those William Wrigley had built on Santa Catalina a generation before his descendants turned almost the whole island into a reserve. It occurs to me that the chewing-gum business might provide us with a model of the various stages along the road that leads from the thoughtless age of the robber barons through synthetic rubbish to the dawn of what we call sustainable production. Even in Swedish shops you can now find chewing-gum with a name that sounds like a joke (Jungle Gum) but that is nevertheless produced from natural raw materials obtained, as in the past, from tropical forests.

And the California condor is also looking for nest sites outside the national parks. In places where respect is the order of the day, they will survive. We saw several of them, flying high. From somewhere or other some lines by the poet Harry Martinson came to mind:

The condor rises high from the Inca's nets and traps
up to where no one disturbs it.
It rises so high that the earth becomes a pill
floating far below.

Harry was a far-sighted man who recognized early on that the earth was just a pill, a view that did not achieve wider diffusion until we saw the colour photographs sent back from the moon voyages of the 1960s. Those photographs quickly formed the basis of the metaphors that coloured the image of the earth held by me and many other people.

Before we had those photographs of a greenish-blue heavenly body floating in space, any threats to the planet could be countered with ancient tales of how utterly enormous the world was. A little destruction in one place could be compensated for by cleaner air and water in another and by endless wild country just a little further away again. The significance of those photographs for the awakening of the 1970s should not be underestimated. They were probably more important than all the words, and, since the threats were huge and the earth merely pill-sized, there were some of us who concluded in the half-light of the forest that the colonial phase of humankind's relationship with nature should be brought to an end. Easier said than done, of course, but, like all utopias, tempting.

That probably explains why, as a matter of principle, I feel ill at ease with reserves, preferring to visit aristocratic parklands, town parks and botanical gardens with cafés or farmland than any of these national parks dedicated to nature and nature alone. I've never been a churchy sort of person, which is one of the few traits I share with Thoreau, whom I've never read with any great enthusiasm. Both he and Emerson (whose garden Thoreau tended) seemed to me to be a touch unctuous in their worship of nature. They are, of course, worthy of great respect as pioneers to whom we should raise our hats, but I find their chastity disturbing. And the good Thoreau does get long-winded: I've read his classic book *Walden; or, Life in the Woods*, first published in 1854, many times, particularly since I myself have spent decades staring out over a lake for days at a time. But, however hard I try, I can't really get beyond the thought that the first fifty pages – that is, the introduction by the translator, Frans G. Bengtsson – is by far the best part.

There is something about Thoreau that brings Oscar Wilde to mind. Not the chastity bit, of course, but in the sense that the stories about these men, embellished with the occasional quotation, are a great deal better than much of what they wrote themselves. Consequently the only quotation from Thoreau I have ever managed to remember off by heart is the short sentence that figures in Bengtsson's account of how it came about that the author chose not to follow in his father's footsteps as a pencil-maker. Thoreau was very skilful and in his youth he helped his father make pencils. The results were superb, and testimonials came from both Boston and New York (where we can assume the real connoisseurs resided) pronouncing these pencils to be first class, at which the young Thoreau ceased making pencils, stating – according to the legend: 'Why should I do twice what I have done perfectly once?'

I'm thinking of stealing that remark – not sure when, though.

And no one hits the nail on the head quite as accurately as Bengtsson when it comes to the question of how Emerson, Thoreau and their less well-known contemporaries in Concord successfully formulated ideas about nature and existence that still hold water today. He writes: 'It never occurred to any of the Transcendentalists to put forward their outlook on life as a "system", for a system always ends up as more or less pretentious humbug which has roughly the same relationship to the underlying psychological realities as a jubilee exhibition has to everyday life in industrial society.' No one could have expressed it better. Their spirituality was independent of any church. They were, we might say, Pietists in the undergrowth. And walkers.

In July 1862, just a couple of months after Thoreau's death at the age of forty-four, the *Atlantic Monthly* published his long essay 'Walking', the first sentence of which reads: 'I wish to speak a word for Nature, for absolute freedom and wildness, as contrasted with a freedom and culture merely civil – to regard man as an inhabitant, or a part and parcel of Nature, rather than a member of society.' It's a fantastic text, meandering and mystical and, indeed, more than a little over the top in places: 'If you are ready to leave father and mother, and brother and sister, and wife and child and friends, and never see them again – if you have paid your debts, and made your will, and settled all your affairs, and are a free man – then you are ready for a walk.' I like it better than *Walden* precisely because the author's constant rambling around the district near his home – at least four hours every day – presupposes a landscape worth walking in. At that time the regions of New England that Thoreau was celebrating – he scarcely ever left the area – were still at a virginal stage of cultivated nature, a mixture of everything that just happened to be, with no other purpose than survival. Unlike the later industrial landscape, which is so harsh and dreary that the walker has to resort to distant reserves while on holiday – a holiday that was something outside ordinary daily life and thus in itself a kind of reserve. But that form of colonialism is on its way out now, or should be. We cannot yet characterize every nature reserve as being a cheap defeat caused by our lack of courage and imagination, but that day will come, believe me. 'Moreover,' Thoreau adds cryptically, 'you must walk like a camel, which is said to be the only beast which ruminates when walking.

When a traveller asked Wordsworth's servant to show him her master's study, she answered, "Here is his library, but his study is out of doors."'

. . .

That the spell should have been broken by a plastic bag, just as it was by Pirsig's empty beer cans in the Crater Lake National Park, will perhaps come as a surprise to some people, since, as we all know, Americans are no great fans of the plastic bag, preferring brown-paper bags that invariably burst halfway between the shop and the Mustang. But that's how it was. A smallish plastic bag from a clothes shop, the size of bag designed to hold stockings or possibly underpants. The company logo, red on white, was so familiar that I didn't react to it until I was on my way back from the viewpoint, and a rock wren, which is another common species in the park, perched close to the bag on the same branch.

Hennes & Mauritz now have outlets at many places in the United States and, indeed, worldwide, though it began as quite a small company in Västerås and central Stockholm selling mainly women's clothes. One business among many. Its name was Hennes, the Swedish for 'Hers', but Erling Persson, who owned it, had bigger plans, and so in the middle of the 1960s he took it into his head to expand into the then newly developed district of Hötorgcity, situated between the Stockholm Concert Hall and Sergels Torg. Hennes already had a shop in the area, but Persson wasn't the sort to be easily satisfied and his eyes fell on the even more favourably located premises of a less fortunate competitor – Mauritz Widforss AB – right on the corner by Hötorget.

Erling Persson bought the lot – lock, stock and barrel. Guns and ammunition were not really up his street, so that part of the business was disposed of over time, but he kept men's clothing. And, funnily enough, he kept half the name of the shop.

Chapter 14

The Girl in Södertälje

When the California Watercolor Society held its annual exhibition in November 1928 – it was in Los Angeles that year – almost sixty painters were represented, each with ten paintings. As was usual at that time, the exhibition took the form of a competition judged by jury.

'And the winner is . . . Gunnar Widforss.'

He had arrived, had finally gained the reputation that would enable him to sell his art and live reasonably well. He had a solo exhibition at Gump's Gallery in San Francisco the following spring and it brought him both money – more than a thousand dollars – and an outstanding critical reception. Lincoln Ellsworth

(1880–1951), the famous polar traveller, and various other rich regular customers commissioned expensive large-scale works, as did the Santa Fe Railway, while art dealers in several towns now stocked his water-colours. Sales were steady. Things were looking good. Stability at last.

Gunnar became an American citizen that summer. He saw it as his duty. His address was Grand Canyon, which is where he celebrated his fiftieth birthday on 21 October 1929. Eight days later the Wall Street Crash occurred. Poor Gunnar.

The Depression did not, perhaps, change everything, but it did change a lot. The crisis was profound and hard and afflicted most people, not least artists. Trying to make a living by painting beautiful landscapes was by no means easy at the start of the 1930s. So we should linger a little while on Gunnar's fiftieth birthday, the high point of his success and before his luck turned.

He was known as Weedy among his friends in the Grand Canyon. They were a motley crew of adventurers, photographers, park rangers, mineral prospectors and others who met to play bridge and poker at the hotels – El Tovar on the South Rim, Grand Canyon Lodge on the North Rim. Or they gathered halfway between, down at Phantom Ranch by the river. In one way or another all of them were lone wolves who came and went, sometimes disappearing for months on mysterious missions, but the park was their fixed point, and no one can survive without friends.

There's not much point in naming their names, as Gunnar constantly does in his letters to his mother. All of them are dead, most of them forgotten. But let me say something about one of these odd characters if for no other reason than to demonstrate that every name you follow up is like footsteps in the sand that lead you on to a world of its own, a world that is full of life and that, in turn, branches out in many directions. There is no end to it. I have traced only a small number of the people who pass through these letters, and it grieves me that so many have been left unnoticed and unremarked. But there is nothing else I can do. Mike Harrison, though, is just one piece of the puzzle among many.

Mike Harrison, born 1897, worked as a ranger and

then later made a career in the Bureau of Indian Affairs. At the time of Gunnar's birthday he and Gunnar had just been on a long trip together, through the big Indian reservation east of the Grand Canyon, in Monument Valley and thereabouts. The country of the Navajo Indians. Mike had friends who lived out there on the plains, and Gunnar, who had never been there, went with him for company and in order to rest and see something new for a couple of weeks after an eventful and hard-working season. That's why Mike Harrison was the only person to know about the approaching birthday. He was told in confidence and he promised not to tell anyone else. Gunnar wanted to avoid a fuss.

The lucky birthday boy got a couple of books from Mike, who had bibliophile inclinations, and letters and telegrams from family and friends in Sweden. 'You are all much too kind to me – I don't deserve all the nice things that have been said in your letters – but it was gratifying anyway – and lovely to hear. I really *celebrated* my birthday – mostly to myself since only one of my American friends knew about it. It gave me the chance to think – and in my thoughts I suppose I spent most of the day over there at home.'

Much, much later, in 1986, when Mike Harrison was almost ninety years old, he remembered the modesty, the considerate manner and kindness at any price, with which Gunnar behaved, even in hopeless situations. It was in a long interview, never published but saved, that he broached the subject of his old friend. He gave a detailed account of their trip to the Indian reservation, talked about their daily life in the Grand Canyon and about Gunnar's relationship with the other painters who came there, and there were many of them. About his kindness: however poorly a colleague's efforts to depict the impossible forms and colour play of the ravines turned out, Mike said, Gunnar would always do his best to find at least a brushstroke here and a brushstroke there that was successful and that he could praise. The old man told the interviewer that in this respect Gunnar was the diametrical opposite to Thomas Moran, his only equal as a painter. Harrison had actually been present in 1924 when Moran, by then an old man on his last visit to the South Rim, had stood looking at a painting by a young, nervous but hopeful painter without saying a word, not a word. Only when he turned and was walking away did any words emerge through his bushy beard: he commented that the frame was well made. Behind him he left an utterly crushed young artist.

I missed Mike Harrison by just a couple of months, no more than that. His mind was clear to the end but, of course, he died eventually, in April 2005 at the age of 107. As I said, he was a book collector.

Gunnar now, working at his little wooden easel, was soon back on the slippery slope to poverty. Come the end he was living in a tent. But the Depression did not slam the door on people on the ground as quickly as it did on the speculators in the stock exchange, so there was still time, even if short, for him to live as the great painter he was. A commission, once again from Lincoln Ellsworth, gave him the chance to make one last visit to Europe: Florence, Venice, Switzerland. And home to Mamma in Ålsten, of course.

In the spring of 1930 Gunnar sailed from Los Angeles to Gothenburg via the Panama Canal in one of the flagships of the Johnson Line. The ship was the *Annie Johnson* and he played golf on deck and swam in the pool. An old story came into my mind. A different painting – earlier. A different voyage, but the same ship.

. . .

'What does the dedication say? Try spelling it out, none of us here can decipher it. But that scribble is definitely Acke's, no doubt of that. But that's all we know.'

As always, the vaguely Chatwin-lookalike expert at the auction house was charming and proper, maybe a little stressed by all the awkward questions about the artworks soon to go under the hammer. A week or so earlier I had spotted an unsigned painting by J. A. G. Acke. Not that I expected to be able to afford it, but still. '*Coastal Scene, Fading into Blue*. Canvas Glued on Panel. With Dedication'. It's always the riddles that snare me, constantly, and the hidden stories. Now I was on the track of something.

'To my friend Sven Hedin. Shared memory of a

run'. With much hesitation and letting my imagination have free rein, I read the artist's flamboyant scrawl. Had the point of his pencil broken? Was he dyslectic or just drunk? Maybe that was it. Acke was well known for his frivolous lifestyle and for some serious partying. The dedication really was illegible, open only to wild guesses. But I was trapped, so I took a chance, bid and bought the painting at auction.

Which is why we had Acke hanging at home. All our friends and acquaintances were invited to decipher the text above the horizon of the sea in the top-right corner of the painting. No one succeeded. Some of the suggestions were interesting. Someone thought that the forename was possibly Siri rather than Sven, someone else wondered more or less in an undertone whether the words might have something to do with mining rather than memory. 'Shared mining,' he muttered, trying it for size. We looked up all the mines in the world with no result. I got hold of everything that had ever been written about Acke – books, exhibition catalogues, articles, and I realized how very unlikely it was that any link with Sven Hedin was going to be found.

They didn't move in the same circles, and they were as unlike one another as two men could be.

Time passed. I had read everything and I had tested hundreds of hypotheses on the wording of the hopeless dedication. I was still on square one when the final book arrived by post one morning from a second-hand bookseller in a country town. It was *By Johnson Ship to Rio*, a small thin booklet about Acke's voyage to South America in 1912. The pieces suddenly fell into place: I recognized the cliff with the flat top in a couple of the paintings that were reproduced. It was undoubtedly Corcovado, the steep mountain in Rio de Janeiro, the one that is now crowned by a gigantic statue of Jesus standing there like a scarecrow.

It emerged that Acke had been suffering from depression and needed to get away. It was his friend Gottfrid Kallstenius who came up with the solution. He had done a number of jobs for the Johnson Line, been commissioned to produce paintings of their ships, and he was on good terms with the ship-owner. So he arranged for Acke to have a cabin on the *Annie Johnson*. Bound for Rio.

And there was more. The booklet quotes the following passage from one of Acke's letters home to his wife: 'Herr Hedén and I raced from the top of Korkuvado – slope 1 in 28 – I won but I've got sore legs . . .' There we have him – Sven Hedén! A completely unknown Swede in Brazil who would certainly have been swallowed up by the mists of history had not his elder brother become famous. Erik Hedén (1875–1925) wrote the first biography of Strindberg and was charged with treason following his involvement in the 1916 Social Democrat Peace Congress that argued for a strike against military mobilization should Sweden decide to enter the war on the German side. A legendary journalist.

That is presumably why the Royal Library acquired part of the Hedén family archive a year or so ago. There was a whispering hush and the faint smell of dust and yellowing paper in the reading room of the manuscript section when I carefully opened the folder containing Sven's letters back to Sweden. A moment later I was transported to Rio, where one fine day the young Hedén receives a visit from a fellow-countryman, who becomes his friend and lodger – the mysterious painter J. A. G. Acke. I carried on reading for the whole of the day.

...

The bold explorer Lincoln Ellsworth had a mountain range in Antarctica named after him. He is a really big hero in the United States – and he was rich. His father, James, had built an empire of coal mines and banks and was awash with money. The old man had his own skyscraper built in Chicago; his coin collection was the envy of the world, as was his library and art collection. He owned, for instance, the big oval Rembrandt, *Portrait of a Man*, that was later donated to the Metropolitan Museum of Art in New York. To lay a hundred thousand dollars on the table was as nothing to him, and that is precisely what he did in 1925, the year of his death, when Lincoln and his Norwegian friend Roald Amundsen were seeking funding for their bid to reach the North Pole by air – an expedition that would soon be in the news.

Lincoln Ellsworth inherited his father's interest in

art and he bought Widforss watercolours in bulk. Property was part of his inheritance, of course, including villas in Florence and a castle in Switzerland. Gunnar was commissioned to paint them, and in May 1930 he reported home from Italy that business was going wonderfully, he had never had better commissions. But there is an undertone of melancholy and homesickness for all that. He writes to Häggart: 'I don't think I shall ever paint in Europe again after this summer.' And to his mother: 'Strange that I don't feel more at home. I long to get back to America or, more accurately, to "the West". Big, open spaces, more light and the weather is more reliable and much finer. Somehow life is simpler there, I can't describe how – but all this hooting of car horns gets on my nerves here. In America we don't hoot our horns endlessly. We don't need to because the traffic is properly organized.'

During the few weeks he was in Sweden the newspapers published an occasional piece about Gunnar. Single-column stuff only, nothing remarkable. 'The official National Park Painter of the USA, the Swede

Gunnar Widforss, is visiting his old homeland for a couple of weeks,' is how a notice in *Stockholms-Tidningen* opens; it does little more than say who he is, that he is currently in Stockholm and, indeed, used to live there, though no one remembers it any more. He passes on greetings from Scandinavian American Artists, a group of painters he was attached to, though no more than loosely; and then there is some rather vague talk about the artistic life on the other side of the Atlantic. *Dagens Nyheter* covered the same interview.

'There is considerable interest in art in America but developments have not moved at the same pace as in Europe. The Modernists are at the stage they were in Sweden a decade or so ago, Mr Widforss says. Young art has not settled down yet and hardly any of it sells . . . Americans, who know what they want in other matters, are rather at a loss when it comes to buying art. They are keen to know that other people have been buying it first. Once an artist has become fashionable, however, he can sell endless amounts.' As if that was only true of the Americans: there is the same hesitation over here, the same fear of getting it wrong and ending up being left on the outside.

Strictly speaking there is only one sentence in any of these notices that caught my interest. Or two, rather. Something he said he missed: 'There's actually only one thing I miss. I should like to take the Swedish dwarf-birch with me.'

What on earth did Gunnar mean by that? Dwarf-birch? I had obviously missed something essential in this case. He had talked about trees on various occasions. Big trees – ancient pines and gigantic oaks, stone pines, palm trees, olive trees, redwoods, aspen and Douglas fir – but never a mention of dwarf-birch. Not until now, at the eleventh hour. He had, of course, painted in the mountains during the war, but mainly in the winter, when all of the dwarf-birches would have been buried under snow. And, if my reading of things was right, whenever he tried to paint during the summer he was driven crazy. A postcard written in July 1918 suggests irritation at least: 'Can't stay in Abisko. The mosquitoes are too bad for a painter and I've been trying to paint using a mosquito net, but it just didn't work at all – couldn't see properly. It was like a veil over everything, which of course it really was.'

Perhaps he just came up with dwarf-birch off the top of his head, just to be nice to the journalist and to live up to what was expected of an emigrant. To be homesick for one plant or another sort of fits with the stereotyped image, and he could hardly say gooseberry, could he? Dwarf-birch is an inspired choice, I think, particularly once you begin to realize that it's a neat evasion and that he is sidestepping something that is no business of the journalists and even less of their readers. Bit by bit in my unmethodical wanderings through Gunnar's life, I had begun to sense that he was missing more than that. Very much more.

Everything points to the fact that Gunnar suspected this would be his last visit to Sweden. His mother, Blenda, was now in her eightieth year, the ranks of his brothers and sisters were beginning to thin out, and he was a stranger to the children of the next generation – they, of course, are the ones who are my present informants. A legend in the family, no doubt, but, when they met him in Ålsten in the late summer of 1930, he was a bird of passage rather than anything else. They still remember Gunnar very well, but none of them can recall a conversation or any closer contact. They met him, that's all.

And so did the girl in Södertälje. It would be unfair to name her name, but I'll tell the story. Even if I have to spend years apologizing for it, I'll still tell it. Not because I'm the one who watches over Gunnar's memory as a friend and therefore feels duty bound to tell what was most important, but because I want to. Because the curtain came down so quickly and then went up again.

Chapter 15

The Great Jigsaw Puzzle Panic

My first thought, a short and hasty one, was that the girl in Södertälje could explain many things, many of the dark corners in the artist's life, the whole of his American adventure even, viewed as an attempt to flee. That she was the piece of the puzzle I needed if I were to see what his motives actually were. But no sooner had I found her than I became unsure – and fearful. I scarcely wanted to admit even to myself that I had been searching for the best part of a year. But I had. I'll tell you about it, but first let me follow Gunnar to the end, because there is a puzzle to complete there, too, in the matter of a craze that broke out at the depths of the Depression and has come to be called the Great Jigsaw Puzzle Panic.

As so often, the whole thing began in America. A toothbrush salesman ordered a cheap cardboard jigsaw puzzle to be used as a free gift to help drum up trade, because the many makes of toothbrush were so similar to one another even then that it was necessary to find some way of outdoing the competition. The jigsaw puzzle proved to be the answer simply because technological advances had led to the mechanical production of puzzles a hundred times faster than had been possible before, when every puzzle had to be sawn out of sheets of plywood (1880–1951). Die-cutting the cardboard on a conveyor belt – using something like a gigantic cookie cutter – was quite extraordinarily cheap. This happened in Queens in 1931 and suddenly toothbrushes found themselves a smash hit.

It didn't take long for other salesmen to recognize that they could actually charge for the puzzles, which cost next to nothing to produce, and in September 1932 what became known as the 'Jig of the Week' was launched in Boston. The idea was that the population needed cheering up, what with the economy having run on to the rocks and mass unemployment being the rule. The customers were pretty well poverty-stricken, but these puzzles – one a week, each with a new theme – cost only 25¢. They were distributed through newspaper vendors and were an immediate success. The first 12,000 sold out in no time at all, no fewer than 100,000 went in November, December went up to 200,000, and in January they easily managed to hit a quarter of a million because by then the craze had spread across the whole continent.

Dozens of jigsaw-puzzle manufacturers now launched themselves into the competition. It was nothing short of a gold rush. The market seemed insatiable. The law of the jungle ruled, and finding the best illustrations became an absolute priority, which was manna from heaven for the hard-pressed artists of the country. We have to bear in mind that the craze for puzzles was a distinctly mass-market enterprise, which meant that a good deal of the art being printed and stamped out by the machines was trash, but, since the expansion of the business was exponential, there was plenty of room for all tastes to be reflected. Buxom blondes, glaring sunsets, sweet puppies and bright-red Indians rubbed shoulders with the real twelve-pointers of art history: Whistler, Millet, Corot, Moran, Monet, van Dyck, Leonardo and, of course, Gunnar Widforss, with a panorama of the Grand Canyon. Apart from Bror Thure de Thulstrup (1848–1930), a cartographer, illustrator of battles and former French Foreign Legionnaire forgotten in his native country, Gunnar was the only artist from Sweden to supply a subject for this inimitable extravaganza.

After six months the great puzzle panic reached its climax in late winter, by which time something like six million puzzles a week were being sold; and then the whole thing came to an end as suddenly as it had started. It was the middle of March 1933. By then the finances of the country were in such an appalling state that the newly elected president, Franklin D. Roosevelt, quite simply closed the banks for a couple of weeks in an attempt to stem the outflow of funds. An unintended side-effect of this emergency measure was a severe shortage of coins, small change, the fuel that kept the puzzle business on the boil. The market crashed overnight and never recovered.

. . .

Gunnar had voted in the presidential election of November 1932. The day before the election he was still hesitating between the two candidates. He thought he could be pretty sure that Roosevelt would win but was nevertheless inclined to vote for Herbert Hoover, the incumbent president, typically enough because he felt sorry for him. The country was hovering on the brink of disaster, no doubt of that, but Hoover had tried as hard as he could, just as Gunnar himself had done. He had battled against a headwind to the best of his ability. Maybe he, too, believed that things would get better and better, when, in fact, they just went from bad to worse.

The Depression was entering its third year and there was still no sign of light at the end of the tunnel. Gunnar was prepared to paint anything at all, unpaid, just for his keep. He worked for a time in Berkeley, painting backdrops for the Chicago World's Fair, a commission from the National Park Service, but one that did not pay anything like what he had been paid a few years before. In order to save money he was living more and more often in a tent, even cooking for himself at times in God only knows what conditions. And he was starting to have problems with his heart – angina, I would guess. He didn't let his mother know much about it, but his letters to Carl Erik Häggart tell in detail how he had overexerted himself hiking in the mountains in October 1933 and been forced to visit a doctor in Phoenix, who prescribed digitalis and later strychnine. The outlook was not bright and yet, during these last years, Gunnar painted some of his most beautiful paintings. He was aware of that himself and it pleased him.

'If I'm allowed to live for many more years and if times get better, I'm sure that one day I'll become famous as a painter of the Grand Canyon, and perhaps I'll earn some real money. That's what ought to happen.'

A couple of years earlier, on his return from Sweden in the autumn of 1930, his income from European commissions had to some extent masked decline. Gunnar got by pretty well at the start. Partly because he briefly experimented with oils, which are traditionally easier to sell than watercolours. One of his large canvases sold for as much as $700, enough to buy a brand-new Ford, but the artist himself was not happy with the result. Painting in oils was simply not his medium and his attempt to do so did not last long.

His watercolours, however, just became better and better – those he finished, that is. All too often he gave up on many of them halfway, losing interest in the subject because everything was taking so much longer now. If the weather was unfavourable a large watercolour might take him three weeks. And then no one could afford to buy it even at half price. 'It's fine to be a painter, anyway, you have so many truly happy hours and tomorrow or the future doesn't bother me much as long as I can paint.'

During these years Gunnar was moving backwards and forwards between the Grand Canyon, Phoenix and Indio in southern California, where he was painting desert scenes. But most of the time he was in the Grand

Canyon, spending long periods down at Phantom Ranch at the bottom of the canyon and at other places close to the river. On one occasion he lived down there for seven weeks at a stretch and found subjects that no painter before him had even come close to. His competitors rarely went beyond the hotels on the South Rim, where in nine cases out of ten they found the view overwhelming. Gunnar, however, was now so profoundly familiar with the landscape and all its moods that he could ignore the bombastic aspects, all the things that encouraged 'pomposity and puffed-up romantic titanism', as Harry Martinson put it in a different context. He preferred to retreat into the woods along the northern edge in order to paint aspens, the subject he loved more than any other. The same trees, every autumn, to the end of his life.

At New Year 1934 Gunnar painted a number of quite exceptional paintings in Death Valley, surrounded by the high California mountains, and, in spite of the fact that poor health prevented his taking lengthy hikes, the following spring and summer in the North Rim's Zion National Park and Kaibab Forest was a successful period artistically, if not economically. His energy and his joy in painting was given a boost by a forthcoming solo exhibition in Saint Louis, Missouri, almost certainly organized by William Scarlett (1883–1973), a man who was the last in the line of Gunnar's really close friends. Scarlett was bishop in Saint Louis and a great admirer of Gunnar's work. Weedy and William, or Bishop Scarlett as he is called in the letters, used to camp together in the Grand Canyon. Another singular character: 'A big strong outdoor fellow. One of the finest men I've met in my life.' Gunnar wrote to his mother and told her that the press called the bishop a Bolshevik because of his commitment to the poorest of the poor, the people hardest hit by the crisis.

There is no mistaking how close they were. It was William Scarlett and not Gunnar himself who eventually wrote to Blenda in Ålsten and told her about Gunnar's problems with his heart. He was the stuff real friends are made of.

The exhibition opened on the last day of October. Gunnar drove the whole way to Saint Louis himself – 1,600 miles to the east, half the continent away.

During the summer, kind fellow that he always was, he had lent his Ford to two teenage boys in Grand Canyon and they had rolled it over and over. They had come out of it amazingly well and, fortunately, the car was insured, so that Gunnar was now driving a brand-new Ford Tudor Sedan Deluxe with a V-8 engine, and the paintings in the trunk were of the very best quality. I'd like to think he was in good spirits.

It was on his return to the South Rim a month later that he died. He had probably pushed himself a bit too hard, what with the stress before the exhibition and the journey home being as long as the journey there. His doctor had told him to avoid the thin air of the Colorado plateau, and, in fact, he had not intended to stay, just to drop off some paintings, pick up a few things and have a game of poker with his friends at the El Tovar Hotel. He arrived, found his friends and they arranged to meet that evening. With just one more errand to do, Gunnar got into his car, started the engine, had a heart attack and died at the wheel at the exit to the hotel. The car crashed into a pine tree. It was straight after lunch on 30 November 1934.

His mother received his last letter, which he had written in Dallas, Texas, on his way home four days earlier.

26/11 Dallas, Texas, Hilton Hotel

Dear, Dear Mamma,

I should have written a long time ago but I was waiting and waiting to see if I could tell you some good news about my exhibition. And since it finished I haven't had any time at all until today.

Even though sales were bad it was still pretty successful. And maybe it will do me some good in the future. I sold three paintings but that was not enough to pay for my outgoings. I'm hoping Saint Louis Art Gallery will buy one (meeting of the acquisitions committee in two weeks' time) as several members of the museum were very interested and suggested I leave a number of paintings behind for their meeting. If they do buy one, all of my outgoings will be more than paid off.

And I'm now on my way back to Arizona. A pretty long round-trip of roughly 3,600 miles. But it's nice to be driving a car when the weather is fine and the roads are good (which they are for the most part). And I got to see quite a bit of the United States. I'm going back to the Grand Canyon first so that the people there will have some of my paintings to sell for me. And I also have to pick up some things I need for the three months or so I intend to stay in Phoenix, where they have the very best winter climate and some quite good things to paint.

When I get to Phoenix (about five or six days from now) and have an address I'll write to you again – it will be just in time to be a Christmas letter.

Bishop Scarlett asked for your address, Mamma, and he told me a few days ago that he had written a letter to you. Enormously kind and thoughtful, but then, the bishop is an exceptionally fine man. And it was a great pleasure to be able to see him every single day for a month. The bishop's friends are really kind people, too, so I've been to a lot of nice dinners.

I know that Bishop Scarlett has told you that my health is not of the best but the doctor is very hopeful that I shall soon be well again. I'm getting on fine, but

I mustn't exert myself at all because the least exertion is both painful and quite dangerous. So I shall have to be careful for two or three months.

Enclosing 2 weeks art criticism – a bit exaggerated, or perhaps very exaggerated, but the lady who wrote it told me that she meant every single word. That was really good!

With my warmest love to all of you,
Your loving Gunnar

A beautiful ending in its way, I thought. Of all the hundreds of letters I read this was one of the finest. It said a lot about Gunnar. Having read my way so far through the correspondence, I knew him. Or thought I did. But there was one more letter.

. . .

I have no intention of saying who wrote it or who it was addressed to – no names at all – but I will just mention that it exists and had existed in the family the whole time. And I'm the first to understand why they wouldn't let me see it at the start. They weren't intending to let anyone in on the secret. Or perhaps the few people in the family who knew about the letter wanted me to reveal my true colours first and, if things turned out well, win their trust. Whichever it was, the story came out anyway, right at the end.

The letter was written on the 24 February 1937 by a woman living in Södertälje and it was about her then sixteen-year-old daughter, her child and Gunnar's child.

Sweden is a small country, full of church registers and national registration lists. If you have access to a name or two, a year of birth and perhaps an address, with a little help you can find out a great deal, sometimes too much. But I didn't hesitate for a moment, not once I had discovered that the girl in Södertälje, born in the summer of 1921, was still alive, had an address and a telephone number and three children of the same sort of age as me. She lived in a small town to the south. I sat with the telephone in my hand for days, not daring to ring. What can you say?

'There's nothing official about it, you know,' she said.

A charming woman. A little surprised, understandably, but of course she would be happy to tell me what little she knew about her father. They had met only on one occasion, it was possibly in 1930, a meeting that had been arranged – that was how she put it. That the artist from America was her father was not, however, something she was told at the time. Her mother had revealed that much later, no doubt out of consideration for her husband, the girl's official father, who was dead by 1930 but had still been alive at the beginning of the 1920s when Gunnar was in Södertälje. A banal story, but a sensitive one: a late child, extramarital into the bargain. And here she was now, telling me about her mother and her four older siblings, all passed away, and that the only thing she had inherited from Gunnar was a stack of paintings depicting the American wilderness.

I gave her the broad outlines of the story of her father. She didn't say much. It was almost all new to her, but she did remember Carl Erik Häggart very well, as a friend of the family. They had lived not far from Astra, she said. That explained something at least. And when I mentioned that Gunnar came from a family with a notable line of artists running through generation after generation, she mentioned that that was something else she had inherited from him. The painter's daughter was herself a painter.

'And your children?' I asked.

'They don't know anything about it,' she said.

After all these years the old lady thought it a bit late in the day to tell her children who their real grandfather was. It hadn't happened before and now she was so old that she didn't want to bring it up. Of course I could understand that. Even I could see that this was a boundary that an outsider like me had no right to cross. It wasn't difficult to respect Gunnar's daughter's wishes on that point. The decision was hers and I decided not to proceed another inch. But it was already too late.

I was contacted just a couple of hours later by a stranger. He introduced himself. I wasn't prepared.

I had of course taken note of Gunnar's grand-children's names and knew who they were, particularly the one who was now on the telephone, since he was a man two years older than me who occupied a prominent and elevated position within his profession. He got to his feet – it sounded like that anyway – when he asked me what I thought I was up to contacting his old mother and asking intrusive questions about the art hanging on her walls. He had been visiting her when I telephoned and, although he had not overheard the conversation, he had seen my name on a slip of paper by the telephone. He had naturally become suspicious, especially as his mother couldn't really explain to him what it was all about, apart from the fact that the conversation had been about Gunnar Widforss, the man who had painted all those old watercolours that had been in the house as long as anyone could remember.

'What is it you are after?'

What was I supposed to say?

I didn't lie. I told him the way it was, partially anyway, told him about Widforss and the trail I'd been following and how it had led by pure chance to his grandmother. A human life, a labyrinth, you know,

I said. You wander here and there, muddle about, try one thing and then another, and suddenly you've hooked something. You can never tell when and you can never tell where. But since his grandmother had clearly been someone who was acquainted with Gunnar in the 1920s and was also part of the circle around Häggart, boss of Astra at the time, I had thought her daughter might perhaps have something to tell me. That is more or less what I said to him. Moreover, and this was just the plain truth, I pointed out that I was doing him a minor good turn by informing him of the value of the watercolours on the American market.

'Increase your contents insurance!' I said.

'Is that what you advise?' he said.

I would have liked nothing better than to tell the truth, the whole truth, and to congratulate him on being related to the only Swedish artist to have a mountain named after him. But there was no way I could do it. What the hell would it look like? So I kept my mouth shut.

Chapter 16

A Fate Worse Than Death

I felt sick. As I said before, who was I to be rooting around in all this?

And now it was too late.

It was no good trying to hide behind naivety, because I'd thought of the grandchildren on the very first day I heard a vague rumour of the girl in Södertälje. From the moment I started searching for the motive for Gunnar's flight, I had foreseen every ethical dilemma, this one included, and carefully balanced it against my curiosity, which, of course, was rock-solid, titillating, provocative. But the notion of flight had been only theoretical, a more or less mechanical fantasy for listless days, a kind of mental embroidery with good gossip potential. Was his creation of the national painting of America the result of a headlong flight from scandal? Could life really be that simple?

The timing certainly matched, in every way. And the attitudes of the period and the dates. The girl was born in the summer, so her mother must have known she was pregnant by late autumn. She must have been more than certain by the middle of December, when Gunnar went off to America, never to return except as a temporary visitor after many long years.

Meaningless speculations. Possibly also unjust. They must actually have decided to keep the paternity secret, to forget it for always. So who am I . . .?

And who was Uncle Ernst?

. . .

During the 1920s my mother, too, lived in one of the fine houses in Ålsten. Only during her very early years, but still. Her father was a lawyer, born in 1889, and her mother, who was some years younger, had studied at the Academy of Art. She rarely painted, though, but took care of her daughters and, as tradition demanded, supported her husband in his career. He was later to become a judge in the district court in Västerås, which is where I remember them being. I certainly shan't forget the grand dinners for the whole extended family that they laid on every autumn in the enormous formal apartment opposite the town hall down by the river. Even the grandchildren wore ties – except me with my bow tie – and we had to try to give an impression of being well-brought-up. This was where my lifelong aversion to organized celebration of any kind began. The housemaids scuttled like ferrets up and down the aisles between the tables.

I loved and admired my mother's parents, as children do. Grandmother was loving and strict at the same time, and grandfather was one of a kind, a giant of a man though little more than a hand's breadth tall. A tiny little fellow, full of stories about anything and everything, which he would narrate in a tone that was both natural and utterly authoritative. Cultured, that was him. The summer I was fifteen years old he despatched me on my moped far and wide around the area to copy all the mossy rune-stones letter by letter, which he then translated for me between the main course and pudding. I wore his praise like medals.

One day grandfather found an old brass box buried in the ground at the timber cottage the family owned at Hästbäck in Bergslagen. It was a late-eighteenth-century tobacco box, the sort that had a perpetual calendar stamped on the lid, which is why grandfather spent his last years here on earth learning everything, absolutely everything, there was to know about the history of Swedish calendar boxes and the various ways they were stamped. By the time he finished there was no one in the whole world who knew more about this limited but fascinating topic. His long article on the subject was not published until 1977 – posthumously.

Well, he did die eventually, of course, at a time when I still hadn't climbed down from the tree, which meant that my interest in matters of inheritance was low. If I remember rightly, the only thing I expressly wanted was a stuffed peacock, which did in fact come to me, probably because everyone else realized how difficult it was to house. I also inherited two other items, which sort of just landed on me without my having asked for them: a sleeping bag and a silver beaker.

The sleeping bag was green and had never been used. I was told that grandfather had acquired it during the Cuba crisis in case the conflict should get out of hand and all hell break loose, necessitating the seventy-year-old judge to sleep out in the forest. I sometimes think about this sleeping bag (now mislaid) and feel a great sense of gratitude that I have never had to experience that level of fear of war. It's a feeling that also helps me to accept and come to terms with the fact that the United States is, for the present and probably for quite some time to come, the world's sole military superpower. When all is said and done, the fragile balance of terror that existed in my childhood had very few advantages.

A peacock and a sleeping bag were the very things for a boy interested in nature, and that was – and remained – my role in the family. But there is no one left now who knows why I was the recipient of the silver beaker. It had belonged to Uncle Ernst, grandmother's little brother, who was born in 1897. He had been given the beaker as a child and it has his name and a date engraved on it. And now, to my amazement when I dug it out, I discovered that my name and date of birth are on it, too, engraved on the bottom. As if it had been decided right from the start that I should be the one to tell his tragic story. The little that there is to tell, that is.

Ernst and my grandmother grew up in the protected surroundings of an affluent middle-class home – their father was an ironmonger in Enköping – and I imagine it was the same kind of domestic environment as that of the Widforss family in Norrtullsgatan in Stockholm. There were fewer children, though, just two girls and him. I know almost nothing about Uncle Ernst apart from the fact that he chose a career in the army, became a second lieutenant stationed at the fort in Boden, where, on 30 November 1920, he shot himself at the age of twenty-three after making a girl pregnant.

My mother has told me that throughout all the years of her youth the story was that Uncle Ernst had died in an accident and that was the only thing that was ever said. There was a portrait of him in the drawing room but no one ever talked about him. It was as if his story were taboo. And no one said anything about the young woman and the girl my grandparents took into their Ålsten home during my mother's childhood. The two were said to be sisters, which is what the girl herself believed. The young woman, who was in fact her mother, kept the secret. The girl was later adopted and disappeared.

In spite of the weight of convention they were kind and decent people and so the girl, who was and is my mother's cousin, came back into the family, admittedly just to the fringes, as a shadow from the past. This was in the 1970s, when she was in her fifties. I have never met her. But since the beaker was quite literally passed down to me, I started digging and asking for answers that scarcely existed. Hesitation, emotional upsets. And I began to think that perhaps all families have their dark corners that ought to be spoken aloud. It's bad enough to put a bullet in your head, but then to be wrapped in a shroud of silence in the name of some sort of merciful oblivion is a fate worse than death. Uncle Ernst was presumably in love, and we know the kind of temporary madness that can imply. Unfortunately, however, he was also an army officer. He knew what had to be done. He had no guardian angel.

. . .

A hundred years ago Mauritz Widforss AB was one of the few gun-shops in Stockholm and it was a business governed by strict regulations. Gun licences, however, were not necessary. Sigfrid Siwertz, whose father worked there, gives the following account: 'The shop eventually moved to Vasagatan, opposite Centralparken. And now and again some unfortunate soul would buy a revolver, go and sit on a bench straight across from the shop window and blow their brains out. And since the gun trade was open and free at that time, it was impossible to refuse to sell them a weapon. But there were times when my father found a way around it.' There can be no doubt that Sigfrid loved his father – he grows the wings of an angel in the story that follows, the story of a girl who came into the shop one spring day.

She wanted to buy a revolver and they let her do so, but Siwertz, at that time the head cashier, had his suspicions. She was a little pale and she was shaking when she told them that bullets weren't so important: she didn't need a box of a hundred, just as long as the gun was loaded. At which the ever more angelic shopkeeper turned his back on her for a while, got hold of a pair of pliers, pulled out the lead bullets and poked in pieces of paper instead. 'His experiment was successful. The girl noticed nothing, paid and staggered out.'

It was, of course, unlucky love. Later on, he heard the whole story from start to finish. 'The girl was a nurse and when one of the young doctors at her hospital was about to get married she lay in wait on the stairs of the bride's parents' home. As the bridal couple emerged on their way to the church she pointed the revolver at her heart, pulled the trigger and – fainted. She regained consciousness in the arms of her mother. A melodramatic tale, indeed, tragic and ridiculous at one and the same time.' Sigfrid Siwertz rounds the story off according to the rules of art by recalling that the unhappy woman recovered her spirits and all in good time entered a happy marriage. They bumped into her on Drottninggatan on one occasion, an elegant woman with husband and children.

'Look,' Sigfrid's father said, 'there's the girl with the blank cartridges.'

. . .

Yes, it's certainly ridiculous. The whole business is banal and trite. There was me, wriggling like a trapped worm and feeling ill for the uniquely silly reason that I happened to know about a long-forgotten affair. Not that the knowledge was making me any wiser. As one more piece of the puzzle, this affair was no more remarkable than any of the others. The hope that all of the pieces would finally fall into place was a vain hope, one that was never fulfilled. Instead, I gave up the theory that Gunnar had fled from an imminent scandal. After all, in 1920 he was over forty years old, which has to be considered a reasonably mature age even for an artist. He probably just wanted to see the world and the rest can be put down to chance.

There is nothing very radical about getting together with a married woman in Södertälje. Especially not for a man who has just decided to travel to Japan and then on around the world, a long way and a long time, taking nothing with him but his paint-box and a recently acquired mantra that states that everything just gets better and better as long as you convince yourself of that fact. Even shy little boys can be irresistible when they are hovering between deciding and departing. The last night is the sweetest of all.

My own fear of hurting anyone subsided bit by bit. The man who'd buttonholed me on the telephone got in touch again, light-headed and cheerful, his head full of ill-considered plans to go to Arizona and climb the hill he had started thinking of as his inheritance. His mother had told him the whole story, which was a weight off my shoulders anyway. Pretty good for Gunnar, too, since there was now some sort of order to his family affairs – if that's what you can call the links forged, possibly by mistake, on that occasion. What do I know? No one can say anything for certain about someone else and the route they follow.

Chapter 17

The Way to Widforss Point

I'll be blunt: Widforss Point is not much of a mountain. It's more of a viewpoint, out on the edge. But the way there is more than worth the effort.

We set off early one morning, Johanna and I, from the small parking area in the forest where the hiking trail – the Widforss Trail – starts, through pines and aspen and against a backdrop of meadows with butterflies, deer and wild turkey, the last acting as living reminders of Benjamin Franklin's uncomfortable thoughts about the cowardice of the bald eagle. These arose during discussions as to which bird would best symbolize the young republic, the United States of

America. Franklin, one of the fathers of the nation, remarked that the eagle was in fact a cowardly scavenger with all-round low morals and consequently unsuitable as a symbol. The turkey, however, was eminently suitable. He pointed out that the bird was combative and brave in defence of its flock and that the mere sight of the colour red enraged it to the extent that it attacked English soldiers, whose uniforms at that time happened to be red. We can imagine what a propaganda triumph that would have been during the Cold War when red fellow-travellers were being hunted high and low.

Benjamin Franklin undoubtedly considered the turkey to be both conceited and a bit stupid – 'vain & silly' as he put it – but its good, upstanding qualities easily outweighed those traits. The thought was an irresistible one, especially to someone like me with a long history of latent anti-Americanism, based on the partly unjust wars waged by the superpower in countries overseas. The politics of the bald eagle. Now, however, after just a couple of short weeks on the Colorado plateau, my feelings inclined more to friendship and admiration for the people on the ground. Sure enough, they can be more than a little bumptious at times and some of them are manifestly vulgar, but what really strikes the visitor is their generosity. Goodwill and the absence of cynicism.

Everyone says that Gunnar, too, was enormously kind. Generous. I know that people say pretty ridiculous things about the dead – obituaries are rarely to be relied on – but there is a tale here of happiness and friendship, one that is hardly likely to have been invented and is therefore likely to be true. Somewhere or the other one of his friends from his last years in the Grand Canyon reported that he and Gunnar's other friends in the park had to be careful about praising Gunnar's best paintings. His masterpieces. The problem was that anyone who said too much about a particularly successful watercolour was likely to be given it as a gift, soon if not at once, and Gunnar really couldn't afford to do that. So they learned to lie low. Poker was, of course, played for money then as now, but every time Gunnar joined them at the table they all exchanged their dollars for Mexican pesos, a much weaker currency that made the stakes seem high when they were in fact small. Deceitful, maybe, but thoughtful, too.

There are times in America when the barbs of justified criticism are camouflaged in a cloak of praise, which in its own way is kindly, though it can sometimes smack of insolence. As was the case recently when a couple of scientists of the entomological sort chose to name two previously unknown varieties of beetle after the president and his secretary of defence in honour, they claimed, of their brave policies in other people's deserts. *Agathidium bushi* and *Agathidium rumsfeldi* are, as every expert knows, millimetre-sized wee beasties at the bottom end of the obscure family of slime mould beetles.

And now that we are at last on our way, on the Widforss Trail, it's perhaps time for us to say something about the names, not the names of the birds along the path – the pygmy nuthatch and the American robin we saw and added to our daisy-chain of memories – but the names of mountains and buildings and rivers, anything that immortalizes the dead and departed. It's so easy to end up with the gnawing fear of being forgotten.

...

The members of the United States Board of Geographical Names knew exactly what they were up to when they decided in the autumn of 1938, four years after Gunnar's death, that a certain point on the North Rim should henceforth be known as Widforss Point. It's quite apparent that the painter's friends, the people who knew him well and knew which paths he liked to take through the woods, had put their knowledge and familiarity to good use. As far as we can judge, for the sake of accessibility, customers and perhaps friends, Gunnar lived most often on the south side of the Grand Canyon, but there can be no doubt about where he felt most at home, where he thrived in the landscape – and that was here, in the woods on the northern side.

During his last four years in particular he had often walked the trail we were now following. He painted the aspens. I had seen some of these paintings before we came to the United States, and, while I thought they were beautiful, I also thought they were a little unreal. There was something about the colours and the light. Now I know. It was early summer and snow was still lying here and there, but the aspens were in leaf and, seen against the light, their foliage and the almost white bark of their trunks sang the same songs as Gunnar's watercolours. Places have their own special light. Maybe the secret lies in the altitude – 8,000 feet – and the amazing clarity of the thin air, although the smog coming up from California makes it hazier now than it was then.

The light and the view, in that order. And with the passing years Gunnar also came to recognize that the foreground is a lot more important than the gorge itself for anyone wanting to paint the Grand Canyon. The trail winds for five miles through ancient forest before it reaches Gunnar's cliff or viewpoint or, at any rate, the point at which you turn back once you've eaten your sandwiches and had your fill of the blue void of the valley and the red of the mountains. The same route back. What is so calming and really makes the experience is that time after time the trail goes so close to the edge that you can catch a glimpse of the abyss through the trees, but not much more than that: you hardly ever come to the dizzying brink of the precipice, not on the Widforss Trail. You will more frequently find yourself walking in the foreground, beneath a vault of ancient pines, or over burned clearings with aspens and wild lupins, or through the lush under-vegetation of damp side-valleys. Rich, varied, nothing to excess. Here and there you will meet people, not many but some, and you'll say 'Hi' and 'How are you?' and 'Fine' and 'How about you?' All in complete harmony, the whole day. Smiles, blisters, drizzle even. And birds everywhere, most of them unidentifiable, but birds. The broad-tailed hummingbirds and bluebirds don't mean much to me, but the bullfinch calling in the air and the three-toed woodpecker that suddenly settled by the base of a massive rotting spruce – not since my days in Muddus! I have to say it quickly because I can't say more. I already feel too much has been said. Must move on at once, on tiptoe, before . . .

It doesn't take more than someone coughing in the room next door for me to flee for my life. Without a word.

. . .

For some reason the transport of Gunnar's effects back to Sweden was delayed several years. Maybe it was easier said than done to trace and gather all the unsold paintings hanging in art shops around the whole of the United States. Then the war came and everything ground to a halt. That's why there wasn't an exhibition in his memory in Stockholm until October 1946, and even then it was held at a framer's rather than in a proper gallery. Something like a hundred paintings were sold off cheap. Where are they now? I imagine they are lying in dark closets and cold box-rooms, or hanging on the flowery wallpaper in remote summer cottages. After all, they are only watercolours, slightly unreal, and by a painter no one remembers.

The Raisin King

Contents

To my mother and father

It was a long story, and like most of the stories in the world, never finished. There was an ending – there always is – but the story went on past the ending – it always does.

Jeanette Winterson

Chapter 1

Rigour

My sewing teacher was a bit of a philosopher. It was 1971. I had just started upper secondary in Västervik, at a school now named after the feminist writer Ellen Key – one of the few truly and justifiably famous celebrities of the district – though it was called something different in those days. A big stone building in the middle of town and a hundred years old even then.

The September air was clear, and the sky cloudless and a chill blue. There was a smell of the sea in the air, as there usually is in Västervik in autumn.

Everything was new and exciting. I was still going around with my pockets full of conkers and the saddle on my bike was maybe that bit lower than anyone else's, but several new school subjects nevertheless pointed to life being about to begin in earnest. We had sewing, for instance, which only the girls had done before, and at the start of term we were going to learn to knit.

The basic principle was reasonably simple, as was the manual skill. There is no great art in knitting a scarf, which is what I had decided on. I discovered that in the very first lesson. I chose a ball of bright yellow wool and set to with all the happy energy that accompanies a newly acquired skill. My fingers were small and had long been used to dealing with the antennae of dried butterflies and microscopic beetles, which is possibly why the scarf was both long and pretty, but as stiff as a board and completely unusable. Very disappointing.

'Rigour,' my handicraft teacher said kindly, 'is laudable, but can easily be taken to extremes.'

I have frequently had reason to remember those words.

Chapter 2

The San Francisco Earthquake

The firestorm that spread from district to district after the San Francisco earthquake on 18 April 1906 destroyed one of the finest natural history collections in America. It represented more than three decades of patient collecting by the Swedish zoologist Gustaf Eisen (1847–1940), a curator at the California Academy of Sciences for many years.

Everything was lost, including his personal possessions. Library, archive, correspondence, the lot. He had to start again, at almost sixty years of age. Admittedly he had done it before, several times, but still.

I have often wondered how he took it. Did he weep?

I don't think so. He wasn't that sort. And anyway, he happened to be on the other side of the globe when the disaster struck. His first encounter with news of the earthquake and the fires came while reading the papers over breakfast in the Bay of Naples. The catastrophe may actually have been something of a liberation. I don't know for sure; I merely have a suspicion. Gustaf Eisen was a man difficult to get close to, even during his lifetime. He went his own way, like a cat. Mysterious and evasive.

'Eisen for dinner,' Strindberg wrote in his *Occult Diary* in the autumn of that year. 'Among other things he said that earthquakes in America are foreshadowed by the arrival of birds. They are white below and black above, resemble waders, but the species is unknown and they are called earthquake birds.'

That was the last time Eisen visited Sweden.

He is one of the strangest men I have ever come across. One of the loneliest, too, perhaps.

The study of earthworms that occupied the first half of his life – the genius of his classification was even admired by Charles Darwin, who wrote and thanked him personally – was not something he wanted to return to. That was all over. His collection was gone and so was his enthusiasm. In my mind's eye I see him rising from his chair in Naples, stretching, looking all round and sort of sniffing the air like an aged bear.

He decided instead to dedicate his life from then on to the study of glass beads. He had already been moving in that direction. His idea was that a knowledge of beads – artefacts that exist at every stage of culture from the Phoenicians onwards – could provide a method of archaeological dating. He stuck at it for ten years. Travelled and travelled tirelessly, visited museums and collectors, and made paintings of all the beads he saw. All of them, everywhere. He was a capable watercolourist.

One fine spring day a hundred years later I found the manuscript:

Glass beads are not just delightful to the eye, fascinating to collect and interesting in general. Studied in the right way they are also extremely interesting to the archaeologist, who, like a modern detective, can weave history and legend out of the scattered and confusing strands that were perhaps initially taken to be insignificant scraps but that in the hands of someone capable of solving their riddles can bring us into close contact with the people whose history we are attempting to trace and comprehend.

I have also seen his watercolours. They have been lying totally forgotten for more than half a century in an archive in Östermalm in Stockholm. All 40,000 of them and they are the prettiest little pictures imaginable. Organized according to a magnificent system, a whole universe in miniature, though I won't comment as to whether it is usable or not. The war intervened and it was never published. Other things intervened, too, and he started again.

Why are people so persistent?

What is it that drives them?

Eisen eventually found the Holy Grail. Literally. The Grail! The 2,000-year-old silver cup that romantics throughout the ages have dreamed of and searched for but never discovered. Eisen, however, was a practical man with a connoisseur's eye, so he found the actual grail, which turned out to come from Antioch in the Roman province of Syria. I've seen it. It's a richly ornamented silver goblet now on display in a very prominent position in the Metropolitan Museum of Art on the edge of the woods in Central Park on Manhattan. It proved easier for Eisen to find a publisher for the book he wrote about it, *The Great Chalice of Antioch*, which appeared in 1923. Two magnificent volumes. The biggest, heaviest and most valuable book I possess. Put legs on them and they would make tables.

The reprint of a shorter version in a smaller format that came out in the 1930s can still be bought, as can Gustaf Eisen's famous book on the cultivation of raisins in California, *The Raisin Industry: A Practical Treatise on the Raisin Grapes, Their History, Culture and Curing*, published in 1890.

Famous among experts, that is.

The multifarious activities of my fellow-countryman Eisen are to all intents and purposes unknown today except by devotees of a small number of esoteric subcultures: Gotland botanists, fig farmers, earthworm taxonomists, students of the Maya, Grail mystics, viniculturalists, historians of national parks, glass experts, alpinists, theosophers, collectors of cylinder seals, raisin farmers, Strindberg experts and diverse other fanatics, including religious bibliophiles and more. As far as I'm aware there is no contact between these groups, each of them having their own Eisen, who frequently exists as little more than a name in a footnote in small print, seen by few and read by fewer. And none of them really know who he was.

Right at the start my feeling was one of fear, followed by depression. Poor fellow. Is that what happened to him? The Holy Grail! Did he really believe in that? Had he gone mad? Or maybe that is precisely what he hadn't done? I put my head in my hands.

But my feeling gradually turned to something resembling joy, a sort of cheerfulness that, to me anyway, is associated with long walks in beautiful scenery. Confidence, perhaps, and rest. The *New York Times* reported on Gustaf Eisen's ninety-third birthday party, held in an enormous apartment on Park Avenue. Folke Bernadotte (the UN peace mediator murdered in

Jerusalem in 1948) was one of the crowd, for heaven's sake, as was the writer Rosalie Edge. Eisen was seen to blow out all the candles on his cake, all in one breath. There's a picture of him and the cake.

The birthday boy raised his glass of champagne and called for silence. He thanked both those present and those absent for their birthday greetings. He even managed to squeeze in his old friend Strindberg even though the latter had been dead for almost thirty years by that point. Typical Strindberg, he manages to poke his nose in everywhere. But Eisen was in fine fettle, and while he had Strindberg on his hook he seized the opportunity to compose a little poetry of his own:

Spurred by the genius of your hand
Your fame sped from land to land.
In dramatic art you led the way
While I through nature's storehouse strayed.
The time must surely soon be nigh
When we shall meet again, you and I,
And earthly life will be no more
Than memories for us of times of yore.

Then, friend, perhaps we two will strive
Once more to solve the mystery of life.

Well, poetry was clearly not his forte, but, for all that, there is something about the last couplet that does hit the bull's eye: 'Then, friend, perhaps we two will strive / Once more to solve the mystery of life.' They are lines that could serve as the preamble to many things in this world. I found them in a yellowed cutting, probably from a Swedish–American newspaper that reported on the party, and I preserve them as a relic.

Eisen passed away a couple of months later.

And since it was Gustaf Eisen who created the Sequoia National Park in the Sierra Nevada in California and thereby saved the biggest trees in the world for posterity, that is where his ashes are interred, right at the foot of a fine mountain, Mount Eisen.

So I was off again. One last journey.

But before that, between striking camp and departure, I took the time to do some research. It wasn't easy to guess which thread to draw on in order to tease out his story and perhaps eventually be in a position to roll it up in a neat ball, but in the end I decided to start where we usually start. I did not, of course, know whether the answer to the riddle really lay in childhood, but that is where I began my search.

Chapter 3

Wonderland

The scent of tansy in full bloom unexpectedly evoked memories of a summer night long ago, memories of a street-lamp on a gravel road that curved through an idyll of houses that slept between forest and sea. In the August darkness a boy has halted under the light and his eyes are following the fluttering movements of a moth around the lamp; its movements are casting shadows on the road and the boy is thinking.

The last thing on his mind is writing an account of things.

He has no desire to be anywhere else and he has almost no history. He is so deep in thought that he fails at first to see a badger approaching along the stone wall that lines the road. Neither of them notices the other until they are very close and then both are equally alarmed. The badger does a sudden turnabout and runs back a short distance before veering off into Dr Colfach's jungle of lilac and spiraea. The hedge, as it was called.

The boy stayed there. Alone in the cone of light, utterly motionless.

A twelve-year-old should never be underestimated.

I willingly allowed myself to be drawn back to that starting point.

I turned twelve that summer, August 1970, so I was getting dangerously close to the borderline when the spell breaks. But I was on the safe side of it for a while longer, free of intentions. I probably had more in common with the badger and with hedgehogs than with friends just a couple of years older than me. I was still myself – there was no loneliness – and I went down to the road by that lamp every night, or the warmer ones anyway, as if escaping into adventures, some of them imaginary and mysterious, of course, though not all of them were. The story of the theft, for instance, was all too real. That, no doubt, is why I had forgotten it. Or, shall we say, suppressed it. I've never told anyone about it; never, not anyone.

This is the first time the memory has risen to the surface. The tansy was growing at the foot of the lamp-post.

Almost everything you do at that age is by way of preparation, more or less fraught with consequence, but in spite of that I still believe that the only words necessary to describe my chase and my dreams then were that I collected butterflies and moths. It's an explanation that has the advantage of being true. My collection was my dearest possession. My recurrent heroic fantasy was that the house we lived in caught fire and forced me to make a choice – with no time to think and hardly a stitch on – about what to save first. The answer was the same every single time: my butterflies and moths.

That's the way it was, no doubt about it, and the night the deed was planned was a really still night. The head gardener's daughter had long since turned off the light in her gable-end room up under the roof ridge of the house closest to the street-lamp; no one ever saw a sign of the old couple behind the oleander hedge even in daytime; and a little further up the hill Kalle Kongo and his old mum still didn't have electricity, which was slightly peculiar but not that unusual on the outskirts of Västervik when I was a child. The Åkerman house next door was unoccupied and owned by someone no one knew or knew anything about: the women referred to him as the Dream Prince because his house was so nice, what with its romantic veranda from which you had a clear view over Kit Colfach's impressive house with its half-wild park-like grounds that sloped down to the water close to the coachworks. The coachworks doesn't exist any longer but at that point – the start of the 1970s – it was still fully operational. Not at night, though, when the iron gates were locked; everything was silent, and the factory lay in total darkness. Sneaking in there was strictly forbidden and was not something I ever dared do.

As far as the lamp was concerned, however, my courage was a little stronger. My plan was to steal it.

The background to that idea was simple and, given the surrealist logic of a child, I would like to think it was perfectly understandable.

. . .

Suddenly, just for a moment, I thought I saw a connection. As if Eisen's story reminded me of something. I had been following him in archives and books for more than a year and had come so close that I sometimes had the feeling that things were the other way round, that it was him chasing me. Anyway, I was quite determined to follow his meandering tracks. It was fun and perhaps there was no more to it than that.

That he reached ninety-three years of age wasn't something I gave much thought to at first. A considerable age, there's no doubt about that, but not that remarkable for a collector and traveller like him. They really do live a long time. Ernst Jünger, who wrote about this interesting phenomenon in his seventies, didn't desert his beetles until he was 102 and there are dedicated entomologists who consider his case to be quite normal.

The figures told me nothing. It was like reading a bird book that tells you that the wingspan of an eagle can be as much as two and a half metres: no one actually realizes what that means until they see the huge bird at close quarters. In Eisen's case it was only when I began digging into his letters, the few that are left, that the full breadth of the life he lived became visible. Generally speaking, his correspondents were unknown to me, but a number of them were people I was familiar with. Charles Darwin had corresponded with Eisen about earthworms. Some correspondents were older still, having been born in the eighteenth century. Others were more recent: when he was sitting like an eagle owl in that enormous apartment in Manhattan he received a letter from Nils Dahlbeck, environmentalist and radio and TV presenter, about peculiar trees in northern Uppland.

Nisse! That jerked me awake, and I was soon lost in thoughts of how long a life can be and – in fortunate cases – how rich. Nisse Dahlbeck is no longer with us, but he and I had a good deal to do with one another back at the start of the 1980s, when I was making a programme for Swedish television called *Field Biologists: Nature and Youth Sweden*, the movement Dahlbeck was involved in starting just after the war. We 'Field Biologists' have never found it difficult to get on with one another, irrespective of age; all of us wander around with a magnifying glass in our pockets.

. . .

Tansy, the smell of windfalls and flowering tansy. Back to that. I've gone in a circle.

From an evolutionary perspective the ability to hide in the vegetation must have been a great advantage in primaeval times. It was probably more significant than either speed or muscular strength. That's why children still play hide-and-seek and are attracted to dark corners where they can sit flicking their low-powered pocket torches on and off. It's a behaviour that is deeply rooted in human biology. Like collecting.

There was, of course, nothing wrong with the butterflies that flew around in the daytime sun, nothing at all, but there weren't that many of them and to all intents and purposes they could all be collected in two summers, perhaps three. There was the odd difficult-to-identify blue or an uncommon mother-of-pearl still to be located in the area, and the Apollo butterfly eluded me for a long time, but these alone were not enough to satisfy me.

While we are on the subject, I'm embarrassed to say that my two Apollo butterflies, a male and a female labelled 11 July 1971, are not specimens I caught myself. They were alive in a big glass jar that smelled of salted gherkins when I bought them from a couple of younger playmates who lived by the road-end, a stone's throw from my house. I had actually gone into commerce in a small way and had contracted to pay two kronor each for any butterflies absent from my collection. I had signed up all the small boys I knew, including these two brothers, sons of a man who enjoyed some degree of local fame. He was known as 'Tajmarn' ('The Timer') for reasons I never fully understood but that were said to have to do with his exploits at the 1936 Berlin Olympics, where he was victorious in a sport that was completely new to the games. Invented in Bavaria, it involved paddling a sort of folding canoe as fast as possible. It was a sport that never featured again. On one occasion I was actually allowed to hold the gold medal in my hand.

The brothers had caught the Apollo butterflies at a place called Åldersbäck, south of the town, and I remember the catch being handed over on the rubbish tip behind the outhouse at my house. I cut a straightforward deal. No haggling, four kronor in cash, the jar thrown in. Embarrassing, I know, but sordid commercialism is one of the basic levels of civilization and only personal experience can make us recognize how hollow it is.

Daylight, then, was too limited. Night, on the other hand, saw a virtually infinite number of species of moths flying around, admittedly many of them small and rather drab, but some of them the size of bats and others wonderfully beautiful. Hawk moths, Bombycids, Geometrids, owlet moths. So I made summer nights my special field. All that was needed was a good lamp and a little imagination. Then you could sit for hours in a sort of nest of light while imagining all the things that were flying around in the darkness and that might perhaps be attracted into the lamplight.

And it did happen. Remarkable moths came my way. The problem was the lamp itself.

Things had got off to a good start the previous summer, which was 1969. Although I knew that the best light sources to attract moths were those that emitted ultraviolet wavelengths, I had nevertheless enjoyed quite amazing success that first summer with an ordinary lamp. The reason was probably because I had been using one of my father's photographic lamps and the sheer power of the light compensated for its poor quality: it was a 500-watt lamp, the size of a saucepan.

I knew that real entomologists had advanced quick-silver lamps of 175 watts at most, but, since they were both expensive and difficult to get hold of, they were beyond my reach.

I mounted the lamp on an old bedsheet on the lawn at home and used to switch it on as soon as it was dusk, thereby illuminating the whole garden from within rather than floodlighting it from without, so to speak. Since we lived in the woods, our garden was surrounded by large oaks and pine trees and the magical effect of this light was to make them seem to lean inwards, as if they were being nosy. The shade behind them was so sharp and black that absolutely everything in it was hidden.

I sometimes imagined how splendid this sight must be from space and I had no doubt that my lamp could be seen from an orbiting rocket. This, after all, was the year the American Apollo project succeeded in landing two men on the moon and the distant view of earth from above was new and irresistible.

Much, much later I recognized the same warm space of pure light in *Wonderland*, one of Strindberg's best paintings, and one he painted when far from home.

Anyway, disaster came in the form of a raindrop, a single raindrop falling from a sky that was virtually clear, as if someone in space were shedding a tear for the boy down there with his ridiculous net and his unrealistic hopes. The big studio lamp exploded with a great plump puff, a sound that is probably unique to an overheated 500-watt lamp. Darkness fell on the garden. It all happened very quickly, simultaneously accompanied by a squall of tinkling ice-crystals, though they were actually the thinnest of thin glass. Millions of fragments, microscopic fragments.

After that my father's willingness to let me borrow his lamps waned. And it does have to be admitted that they weren't exactly cheap.

All of which explains why I ended up under the street-lamp.

I had learned from an older friend in the Field Biologists that the lighting that lined our streets and roads consisted of what were called mixed-light lamps and that they gave out sufficient ultraviolet light to attract even very choosy moths. All you needed, I was told, was a long ladder and a large oven-glove, because stealing them by day was not to be recommended and at night the bulbs would be extremely hot. Nor was there any point in leaving the bulbs where they were and standing underneath the lamp, because then the moths would be out of reach.

I had already made one attempt early in the summer. Not with a ladder, because the only one we had was of the very short variety. I used an invention of my own. Well, not exactly of my own. I had watched what the men from the highways department did when it was necessary to change the bulb in a street-lamp: they had a long pole on the end of which was a rubber cone that fitted over the bulb. It was then just a case of turning the pole and unscrewing the bulb. Child's play, or that's what it looked like.

Getting hold of a pole six metres long wasn't difficult. This was before the days of communal sewage works, so there were many places in the woods where individual sewage outlets manured the earth so richly that aspen saplings shot up like pipe-cleaners. They were long and not too heavy once the bark and side-shoots had been trimmed off. The rubber cone was trickier, but I settled in the end for a discarded plastic toy – a cone-shaped piece of plastic mesh, part of a gadget that enabled you to fire a ping-pong ball by pulling a trigger. These things were sold in pairs and the idea was that the other player should catch the ball in his cone and fire it back. It wasn't a game anyone played for very long.

I lined the inside of the plastic cone with rubber from an old bicycle inner-tube and fixed it to the end of the aspen rod with wire and insulating tape. Perfect. The bulb would soon be mine. After hiding my equipment in the woods I waited for the first suitable opportunity, which must have come quite early in the summer, since I remember a nightingale singing lustily in the hedge as I committed the crime. Or attempted to, because things didn't exactly go to plan.

What I hadn't taken into account, though I was aware of the fact, was that mixed-light bulbs are very sensitive to knocks. You only have to kick the lamppost and the light goes out. They come on again after a while, but that wasn't going to be much help in this case. The moment I managed to fit the cone around the bulb everything around me suddenly went darker than ever before, all the more so since I'd been dazzled by standing staring at the light while trying to control the aspen rod, which had a tendency to sway from side to side. At least no one can see me now, I thought, while carefully twisting the rod anti-clockwise without being able to see what I was doing. Nothing happened.

I put a bit more effort into it and suddenly it moved easily. Too easily. The plastic cone had come off the end of the wooden rod, and when the bulb came on again a few moments later I could see that the cone was stuck up there. The smell of burning rubber began to mix with the scent of lilacs and I went home.

Fortunately, the cone had fallen off by the following morning and the bulb itself proved to be surprisingly robust. The bits of rubber on the glass soon melted off and everything went back to the way it had been before. I was now the owner of an unusually long pole and I toyed with the thought of putting a net on the end of it in order to deal with high-flying moths, but I rejected the idea. It would have looked ridiculous and been humiliating. I was very sensitive about things like that, even in those days.

. . .

In the end all that stays with you are the smells. And perhaps the clink of empty bottles as the brewer loads them on to the delivery truck, or the sound of the hot-bulb engines in the fishing boats out on the waters of Tre Bröders Sund. That kind of thing. These days the only memories I rely on absolutely are those that have never been documented. Not because photographs lie, but because they banish the half-lies that make up the authentic half of all true memories.

. . .

Later the same summer my quota of humiliations reached a level that sufficed for some time to come. I had read something interesting in Torben Langer's *Allhems Butterfly Book*, a magnificent volume with beautiful pictures that I used to borrow from the town library. There was one passage in particular that stirred my imagination. It was just a single sentence: 'It is also a well-known fact that it is possible to attract large numbers of hawk moths to pears, since they emit powerful ultraviolet light that is invisible to the collector.'

It sounded a bit mysterious but did not seem completely unlikely – after all, as I had read in other books, insects seek all kinds of half-rotten fruit and I knew how effective invisible UV light could be for collecting moths at night. So here we had what was obviously the perfect combination, pears were commonly available and it was just a matter of finding the right sort.

It was not until it was too late that I realized that the explanation for this strange statement was both simple and silly: in Danish, the original language of the butterfly book, the same word is used for 'pear' and for 'electric bulb', so the statement was just an elementary translation error made when the butterfly text was being translated. How could I have known? Nature is full of wonders and I was prepared to accept all of them. Your friends might play tricks on you, but books wouldn't. That's what I thought. If glow-worms can emit visible light – which is, in itself, amazing – there seemed to be no reason why pears should not be able to squeeze out a little invisible light.

That, I imagine, is how I put two and two together. The mistake, however, would have been easier to live with afterwards if I hadn't worked up the courage on the way home from school one day to approach one of the old women who ran a fruit-stall on the square at Fiskartorget.

'Have you got the sort of pears that emit ultraviolet light?'

I had carefully gone through various strategies before finally deciding to use the word 'emit' exactly as in the book, but I could hear for myself how stupid it sounded and my stomach churned. The old woman, whose diminutive figure was bundled in shawls, rags and woolly cardigans, just glared at me. There was a smell of smoked eel in the air. I felt crushed and didn't dare guess at the thoughts that were stirring beneath her headscarf, but at that date I don't imagine knowledge of ultraviolet waves was particularly widespread among the stall-holders on Fiskartorget in Västervik. She gave a snort, no more than that, before a more reasonable request from another customer saved the situation and gave me a chance to retreat.

I'm not sure who it was who cleared up the misunderstanding, my teacher perhaps, but I do remember the hearty and spontaneous outburst of laughter and the mortification it caused. I blushed and cringed and determined never again to involve adults in my efforts to catch moths at night. From now on I'd take care of myself.

That very evening found me back under the street-lamp. A gigantic owlet moth was circling the lamp, round and round and round, without ever coming within range of my net. I could see that it was a blue underwing. I don't know how long I stayed there but the tansy was in flower and the smell of the tar impregnated in the wooden lamp-post hung like a vague veil around everything.

The show lasted for perhaps an hour. The moth reminded me a little of a bat. A badger came past. Everything else was still. That was when I concocted my plan.

Chapter 4

The Recluse

Sweden is a small country. Sooner or later we all bump into one another. I actually came across Gustaf Eisen twice, but our first encounter passed unnoticed. On that occasion knowledge of the man who was the model for Strindberg's recluse was quite simply too exclusive, vouchsafed only to professors of literature and dry-as-puffballs Strindberg experts.

I had decided to extend my charting of the natural history of the summer night to include the romantic reflections of the pleasures and mysteries of these nights that is to be found throughout Nordic literature. And since the scent of the lesser butterfly orchid and the call of the corncrake in the half-grown rye are more or less omnipresent in this literary landscape, I soon recognized that I needed to set strict limits if I were to have any chance at all. I focused on Frans Eemil Sillanpää, Sven Rosendahl and Sten Selander, and devoted a whole winter to reading Gunnar Ekelöf. Strindberg, too, of course, and one of the things I listed in my field notes was his short story 'The Recluse' in the volume *From Fjärdingen and Svartbäcken*, a collection of stories of student life in Uppsala.

That particular piece had actually caught my attention earlier, during a year when I had been trying to find a way around my embarrassing (as I thought) lack of interest in literature by studying how Swedish authors throughout the ages have chosen to depict entomologists. The recluse – the main character of the story – is a collector of insects, as we discover right at the start.

> *He never sought anyone's company, was never seen out and was difficult to meet. He lived by the churchyard, in two rooms, probably with a kitchen, too, though no one knew that for sure.*

On his hall door, which had a frosted-glass panel, there was an unambiguous notice: 'Available only until 7 a.m.'

If anyone rang, a completely different door in the hall would open and an ill-tempered elderly woman would stick her head out and ask 'Who is it?' If one failed to respond promptly and politely with one's name and student nation, the door would shut permanently. If one persisted in ringing, the bell would soon fall silent either of its own accord or someone else's.

If one did gain admittance, however, the first thing to be seen was a natural history collection consisting of herbaria, stuffed birds, drawers of insects and aquaria. Next came a study, which resembled a drawing room with traces of decaying luxury. Red plush easy chairs, embroidered mats, threadbare cushions with tapestry edging, oil paintings on the walls – including one by a well-known old Flemish artist – a fine library, a valuable microscope and an easel.

That's how the story opens. The man then described is a rather misanthropic 24-year-old, with private means but no living parents, who devotes his time to zoological studies. The author spends some time describing how he spies on this peculiar student through the window and manages to catch a glimpse of him feeding a medusa in a saltwater aquarium, complete with green algae of the genus *Enteromorpha*. He is a mysterious figure, unwilling to allow anyone to come too close, and he is described as an egoist. Unapproachable, misunderstood. 'I wanted to get to know him better, but I did not succeed. I wanted to hear his story, but he never told it to me.'

It isn't a long short story, just five pages, but nevertheless it is sufficiently comprehensive and revealing, particularly once we realize that the individual depicted in the story is the young Gustaf Eisen. It is possible to prove that several of the events narrated in the story did actually happen during the years he shared with Strindberg at Uppsala University. The latter was involved in the drinking session at Eisen's place that is mentioned in one passage, and, interestingly, he was even allowed to take part in some of the nocturnal excursions around which simpler minds in student circles tended to weave mysteries:

One dark night a friend had seen him creeping into the churchyard with a lantern in his hand. Dreadful things were hinted in a whisper. I defended him but I didn't really know what to believe, knowing him to be a fanatical naturalist. For the sake of his good name I thought I should ask him to tell me the truth of the situation.

He became angry and refused to answer!

And there was another mysterious story doing the rounds at the same time. A young student, diligent and with a real aptitude for his studies, was preparing to leave Uppsala because he lacked the means to continue. He was setting out to meet whatever fate might bring him in the capital. At the very moment of departure the postman came with a registered letter containing both a sum of money and the promise of a similar sum every month as long as the student continued his studies in the same industrious manner.

The handwriting was disguised and the signature false. The story was passed round and eventually became hackneyed, while the fortunate young student corres-ponded with his unknown and pseudonymous benefactor via poste restante.

At the end of the story the recluse with the microscope and the aquaria, far from being the grave-robber the boys had whispered about, is revealed as being a man with the noblest of all aspirations. He says: 'My visits to the graveyard by lantern light had to do with a rare earthworm I have been describing for the Academy of Science Report.'

I was fortunate enough to have the report up in the loft. Five grocery bags filled with an old job-lot of that excellent series *Survey of the Proceedings of the Royal Academy of Science*, an unbroken run of whole years from the second half of the nineteenth century. A gold-mine. In a fit of impatience and anticipation I raised the ladder to the trapdoor leading to the loft. We'll have reason later to return to what I found.

First, though, a few words about the poor student and his unknown benefactor. The former is a fairly realistic portrait of Strindberg himself, who was usually broke, and the latter a portrait of Gustaf Eisen, who in his youth got into the habit – a rather foolish one, at that – of giving away his money to spendthrift friends. The various transactions are well documented.

For what seems to have been rather more than a year in the early 1870s, Strindberg regularly received twenty-five kronor a month from an unknown benefactor who called himself Alessandro Florelli. That pseudonym concealed Eisen and Georg Törnquist, an actor who later became well known – Törnquist was part of the same circle. A number of the letters that passed between them have survived, including Strindberg's courteous thank you letters to Florelli and his later rather more heartfelt letters direct to Eisen after he had worked out where the money was actually coming from.

. . .

Gustaf Eisen was an orphan. While still in his teens he had lost both parents; a cruel blow indeed.

He was born in Stockholm in 1847, a late addition to the family of Frans Eisen, a wholesale merchant born in 1796. They lived on Regeringsgatan and then later in a large house at Österlånggatan 1 in the Old Town, right opposite the royal palace. The mother of Gustaf's seven siblings had died quite a long time before, and his father Frans had married again, this time to Amalia Markander, a woman considerably younger than himself. She gave birth to Gustaf at a time when the older family was already beginning to leave the nest. In the years around 1850 three of his adult half-brothers emigrated to America. A fourth had become a seaman, but he fell overboard and disappeared in the Mediterranean in 1852. His name was Åke and Gustaf remembered him all his life. 'My brother Åke and I were very attached to one another even though he was thirteen years older than me. My last memory of him is of me sitting on his lap with my arms around his neck. That must have been in 1851 and I couldn't have been more than four years old.'

The family originally came from Danzig. The ancestor who moved to Stockholm at some point in the middle of the eighteenth century was Johan Jacob Eisenbletter, an enterprising cabinet-maker whose Rococo furniture with its bulging curves still rouses interest at quality auctions.

Gustaf really deserved a better start in life. He learned to read at a very early age, but he was plagued by illness. As well as chronic bronchitis he suffered from a dislocating hip, food poisoning, measles, scarlet fever, meningitis and so on, seemingly endlessly. In short, he was frail and was given little chance of surviving. In spite of all that, he was sent to Klara Elementary School, an institution notorious at the time for the physical brutality of its regime. It didn't improve his health. He remained there for his first two years at school but he didn't learn much, mainly because he was often confined to bed, on one occasion for nine months at a stretch.

The boy loved his mother. His father, however, is said to have been very strict.

Fortunately, the business was going well, so his parents decided to send him away. A change of air – anywhere, as long as it wasn't Stockholm. And where could you find air fresher than in Visby on Gotland? No sooner said than done. Along with a nanny the boy was shipped off at the age of eleven and he did not move back home until five years later, now fit as a fiddle and with a burning interest in natural history, archaeology and art. Visby Grammar School seems to have been heaven on earth.

In all my days I have never read a Swedish author who praises his schoolteachers to the extent that Eisen does in the autobiographical sketches he wrote in his nineties. Perhaps he hoped that someone would eventually find them at the bottom of an unsorted cardboard box in a library storeroom. Bear in mind that what we are talking here about was an absolutely ordinary provincial grammar school in the period shortly before the last famine years, when the common people of Sweden were quite literally starving.

He mentions four teachers by name, all of them born on Gotland at the start of the nineteenth century.

Gustaf Lindström, who taught physics and chemistry, was the leading expert in the country on the fossil fauna of sedimentary rock types; he was a tireless leader of field excursions and he translated books in his spare time, including Charles Lyell's *Principles of Geology*. He worked at the grammar school in Visby for twenty years before being called to the Natural History Museum in Stockholm as professor and curator of the palaeozoological department.

Eisen's botany teacher was Oscar Westöö. He went down in history as the driving force behind the Society of Bathing Friends and the founder of its botanical garden in Visby. The boys loved him, and during the summer half-year they went out botanizing more or less daily. Then there was Karl Johan Bergman, who was in charge of the teaching of philosophy and history: in time, he, too, made something of a name for himself as a member of the Swedish parliament and as an author.

I think, however, that it was a fourth teacher, P. A. Säve, who had most influence on Eisen and on his many and varied pursuits. With the benefit of hindsight he emerges as the most important of them all.

Per Arvid Säve (1811–87) was one of these passionately romantic cultural historians who dedicated his life to collecting and documenting ancient monuments and folk-memories of all kinds. Nothing was beyond his interest as long as it came from Gotland and was of some age. Folk-tales, songs, dialects, place-names, handicrafts, customs and farming methods. And, of course, rune stones, churches, farms and everything that had been ploughed up by some yeoman working the earth, which, as you can imagine, was quite a lot in the case of Gotland: the number of Viking Age silver hoards alone now exceeds 750.

Säve wrote and published numerous works about his beloved island. He was also the founder of Gotlands Fornsal, the museum that is now the biggest tourist attraction on the island. If any further proof of his restless curiosity and almost manic activity is needed, we could point to the fact that he was the first man in Sweden to argue publicly for nature conservation. It was claimed for many years that Adolf Erik Nordenskiöld, whose Arctic exploration of the North-east Passage made him a folk-hero, was the pioneer, but at the beginning of the twentieth century, when conservation on the German model was beginning to gain ground, it was discovered quite by chance that the incomparable P. A. Säve had been arguing its case as early as 1877 in his far-sighted but forgotten pamphlet *The Last Couple to Leave*.

Four teachers. There was also a fifth, whom Eisen never forgot, but I shall leave him until later. I wouldn't have spent so much time on the teachers at Visby Grammar School if I hadn't been convinced that they played a decisive role in the path this lonely child took through life. There is, moreover, the fact that even after 150 years they remain quite outstanding men and worthy of mention.

. . .

I don't remember much about my own secondary-school teachers and I don't think any of them live on in reference books. The only thing I can remember about my biology teacher, for instance, is his breath. The woman teacher who taught us sewing was, I suppose, unforgettable in some respects, but even then I can only remember the odd bit here and there, as elusive as wood shavings.

Oh, and we did have woodwork, too, during the spring term.

The teacher was known as Woodwork Hitler.

His name was, of course, quite undeserved, but many woodwork teachers were called that at the time, not just in Västervik but all over Sweden. I discovered this from the other boys I met at a gymnastics camp in Malmköping to which I had been deported for the summer so that my parents could have some peace and quiet. Perhaps they were called that all over the world, I thought. Woodwork Hitlers. That autumn I put my hypothesis to the test by asking my big sister, who was studying French, for a translation. The result was depressing. My perceptions of the French were still vague at that point, but there was no way anyone could have been called that, not even a woodwork teacher.

I don't know whether research into nicknames exists as an academic subject, but, assuming it does, it must be a rewarding field. It seems likely that even this little thicket in the vast jungle of language would be subject to the laws of evolution. The thought struck me one day when our own children came home from school, which on this island is a sort of experimental chamber for the kind of rural-life romantics who are prepared to present every imperfection as a gift, or at least as a unique pedagogical opportunity.

This resulted in children being shipped hither and thither from one island in the Stockholm Archipelago to another: that particular year there was obviously a shortage of pupils out on Sandhamn, so they had to travel out there one day a week. Apart from gymnastics they also did handicrafts out there in the outer islands. Since this piece of idiocy happened to coincide with the Kuwait War, during which Saddam came to personify the archetype of evil, it was no more than a couple of days before the unfortunate handicrafts teacher was given the nickname Sandhamn Hussein.

Oh, well, even Woodwork Hitler eventually had some significance for my mental development. On the night I finally decided to steal a lamp I came up with a really good plan and the woodwork teacher was going to be my obedient tool.

Chapter 5

The White-backed Woodpecker on the Island of Gotska Sandön

Strindberg and Eisen got to know one another when they were pupils at Klara Elementary School, and on Gustaf's return from Gotland they became schoolfriends again at the Stockholm Lyceum, a private secondary school. From there they moved on to the University of Uppsala. Their friendship survived into the 1900s, which, given Strindberg's usual pattern, has to be seen as rather exceptional.

An even closer friend of Eisen, however, was Anton Stuxberg (1849–1902), later a member of Nordenskiöld's famous *Vega* expedition through the Northeast Passage and later again curator of the Gothenburg Natural History Museum. He was from Gotland, had grown up on the Stux estate near Fårösund and become Eisen's best friend during his boyhood years in Visby. A happy and carefree vagabond spirit who worked himself to an early death, he was an expert on millipedes, fish and marine molluscs.

In 1868 they made their joint debut as authors with *A Contribution to the Knowledge of Gotska Sandön*. It is the very first dissertation to deal with the island, all sand and dark green pine woods, inaccessible and out beyond the horizon in the middle of the Baltic. It was the result of an expedition the previous summer, when Eisen was nineteen and Stuxberg eighteen. Just boys, then, in spite of which it was published by the Royal Academy of Science. Things were becoming serious.

. . .

An island, isolated and mysterious. It was no accident that they decided to study Sandön, the island over the northern horizon. The whole island was an adventure in itself and, above all, it had not been studied before.

It's clear from Eisen's surviving letters that the trip was being discussed as early as the autumn of 1865, when Anton was only sixteen years old. Eisen wrote to his friend in Visby, told him his plan and asked, 'Have you studied Darwin's hypothesis, Stuxberg?'

The answering letters no longer exist, having been swallowed up by the earthquake, but it is still possible to read between the lines that Stuxberg let Eisen know that he was familiar with his Darwin and that he was rather put out that he had been addressed formally by his surname: they were, after all, old schoolfriends, so he could have used the informal *du* form, couldn't he?

And that's exactly what happened – very quickly. Within six months the tone of the letters becomes intimate and sometimes quite painfully revealing. Longings, fears, dreams. Less than a week after his father's death, in May 1866, Eisen writes how profoundly happy it makes him to have someone to confide in, and time after time over the years he asks Stuxberg to burn these letters, 179 of them in all, the last dated 1881. But Stuxberg kept them, in spite of the fact that the content is sensitive and by no means always to his own advantage. There are harsh words at times, about deceit and broken friendship.

'If you should find him murdered one day, then you can go to the authorities and report me.' These words were written early one morning in November 1871, when they were both studying in Uppsala, and they refer to another student whom Stuxberg clearly preferred to spend his time with. He was given to living it up even at that stage. Eisen tried to protect him, to save him and, indeed, himself, for he must have been suffering from great loneliness. There are times when he nags Stuxberg about his studies and his postponed exams like a worried parent, but he still forgives everything. And he has plenty of money for his friend to draw on.

Lonely, hard-working and with an inheritance to squander, Eisen became a sort of patron, an older uncle responsible for a badly behaved young man, who in reality is only eighteen months younger than him and the same age as Strindberg, whose name pops up here and there in the correspondence. It is held by Gothenburg University and I've read it all. I'm not, however, given to spreading unnecessary gossip: Stuxberg really should have burned the letters.

But they do give a very fine picture of two boys on their way into the world of scientific discovery. At the start they correspond about their collections, about birds' eggs and snails and living creatures in saltwater aquariums, but pretty soon they begin to specialize, Eisen on earthworms and Stuxberg on millipedes. They were still at grammar school when they set off for Gotska Sandön at midsummer 1867. And, had it not been for Gustaf's mother being in poor health, I would have sworn that life was smiling on them. Biological field studies have always provided a wonderful cloak for all kinds of other things. For a couple of passionate and intense years Anton Stuxberg was Gustaf Eisen's very best friend.

. . .

I, too, had a very close friend like that during my teenage years. Thorbjörn Stärner (1958–94). Peace be with his memory.

When my mother rang me on the island one oppressively hot summer's day and told me that Thorbjörn was dead – an ordinary car accident – my whole world crashed down. It surprised me at first. Contact between us had become infrequent over later years. Both of us had married and had children. Fate had taken us in different directions.

Not until he was gone did I understand the full scope of our friendship. It was the saddest day of my life. During the three years we spent in the sixth form in Västervik nothing could part us. Nothing. We did everything together. Fished, photographed birds, played the guitar and partied, trapped moths and chatted up girls, of course. And we travelled to the Camargue and Greece, Spain and Poland, and whatever we did, always and everywhere, we talked and talked about life. Incessantly.

We must have looked an ill-matched pair. Thorbjörn was six foot six and suffered from a lack of self-confidence. I was no more than a hand's breadth in height and generally regarded as stuck-up and annoying. I can remember the nights we sat in my room until dawn was breaking, talking quite specifically about what it is that forms an individual's disposition and what part of it one can shape and govern for oneself.

How we quarrelled! Anything is possible was the line I took. You are wrong was his. But we always made up and by the next day we were off on new adventures. When I heard of his death I just sat alone at my desk in my house on the island, turned my face to the wall and cried inconsolably.

Suddenly I heard the pitter-patter of children's feet. I turned round and met the eyes of our eldest son, seven years old at the time. He had his little sister in tow. He held out his clenched fist, slowly and sort of expectantly, and then opened it carefully:

'Daddy, look what we found.'

In his hand was a beetle, a blood-red *Ampedus sanguineus*, one of the most beautiful of all beetles. The children had never seen me crying before. The atmosphere in the house had been dull and heavy that day, so they had kept out of the way. But eventually it had got too much for them and their concern had led them to scuttle off to a rotten tree-stump and root around until they found a beetle. And now they were standing there with shy wonderment in their big round eyes.

'You can have it.'

'Are you a bit happy now, Daddy?'

Even now I'm likely to start crying at the sight of an *Ampedus sanguineus*, whether from sorrow or happiness I don't really know. A mixture of the two probably. Fortunately the species is common on the island, which has an abundance of the stumps of pines and other trees, all in various stages of quiet decay. Wood-eating beetles are interesting creatures.

. . .

A Contribution to the Knowledge of Gotska Sandön is an apprentice piece, not very remarkable but charming in its dry fogeyishness. 'We are well aware that our collections are incomplete, in spite of which we have nevertheless permitted ourselves to share them, partly so that they may form a basis for future studies and partly so that our material should not remain wholly unused.'

Sticking to all the rules of the genre they begin their

dissertation with a panorama, a general description of the geology, climate and natural environment of the island. I wandered around Sandön myself one summer and, although it was long after their time, their descriptions felt familiar. The only things that were more numerous in their day than in mine were the remains of ships that had been wrecked and 'washed ashore in truly chaotic confusion'. The monotonous wilderness of the pinewoods seemed very familiar to me, as did the shifting sand-dunes and the huge thickets of hazel and the yew trees. Listening carefully between the lines you can actually hear the nightjar's churring voice further inland and the droning buzz of long-horned beetles flying round the lighthouse.

These are not things you forget.

Ant-lion lacewings dance above the sand during the light nights.

Naturally enough it's the lists of species – the score-sheets, so to speak – that have the most enduring value in publications of this kind. The landscape is changing all the time, and the flora and fauna reflect what is happening. In the 1860s people still remembered the reindeer that had been on Gotska Sandön for ages, having been introduced by some optimistic settler. Sidén, the lighthouse-keeper, gave the boys an account of how and when the last reindeer was shot. They entered it in the list of species anyway: after all, they were interested in natural history.

They list everything they see and hear. They find 150 different species of plants, which they think is rather a lot, but they go on to point out that the number would probably have only been half that if the island had remained uninhabited. They found only small stinging nettle, lady's bedstraw, dyer's chamomile, wormwood and many other plants around Nybygget, the deserted farm in the southwest of the island. These flowers had no doubt arrived along with the cultivation of the farm and had then settled in; they are certainly not native species.

That was no great problem, however, not given the way these things were viewed at that time. Rather the reverse: the flora and fauna were enriched by immigration. And that really is the case: I'm not denying that alien species can sometimes be a pest, but much more often they bring us joy and are useful, even if only by making the landscape prettier. But I don't want to get involved in that.

If I did, however, we could talk for hours about the topic on the basis of Eisen and Stuxberg's second book, *Gotlands Fanerogamer och Thallogamer, med fyndorter för de sällsyntare* (1869) (*The Phanerogams and Thallogams of Gotland, Including the Location of the Less Common*). The book is a very detailed catalogue of all the 957 species then known on Gotland. Pedantic systematics, of course, but also a map of the boys' childhood, with the words all in Latin. The result of what must have been hundreds of excursions dressed in short trousers and carrying their vascula. A brew so concentrated that you risk choking on it unless you water it down with five parts of imagination and your own memories.

But let's go back to their debut work on Gotska Sandön.

A list of species can often tell us a lot about a place, but it sometimes tells us more about its compilers in their crumpled tent among the pine trees. Out on Sandön the boys were on the lookout for more than just plants: the abundance of centipedes, earthworms and molluscs in their material bears witness to the fact that their future interests were already forming. But there are other things even more interesting, and several of the discoveries they made that summer of 1867 have continued to be discussed over the years.

The best known is the white-backed woodpecker. They admit that the species is rare, but their sighting has come to be regarded as doubtful, since no one since has seen any sign of a white-backed woodpecker on the island. As ornithologists, Eisen and Stuxberg are remembered for a bird that is assumed to be a false identification, which is sad. Experts mutter sceptically and suggest that what the boys actually saw was a great spotted woodpecker, which is so common on the island as to be nothing worth writing home about. But there are three things that actually point to their identification being correct.

In the first place, they actually include the great spotted woodpecker and list it as a common species there: 'Very Common'. So they knew what they were comparing it with, and the difference between the two species is great enough for anyone with eyes to see, even without binoculars. Second, we can show that the white-backed woodpecker was more common at that time – more or less like the European roller, which, incidentally, is a species they also list, along with the comment 'Rare'. For reasons that aren't clear, the last species has disappeared totally from Sweden now and the white-backed woodpecker seems to be going the same way. Silviculture is what usually gets the blame and there is probably an element of truth in that.

The third link in the chain has to do with the woodpecker's movements. As every bird-lover knows, if you want to find the white-backed woodpecker you need to look in lush, overgrown woods of old aspen trees and pussy willow and fine deciduous trees. Dry and dusty pine-woods growing on sand are entirely the wrong environment. All fine and good, but . . . but it just so happens that the white-backed woodpecker is one of those species that sometimes sets off on excursions in large flocks. In the autumn of 2008, for instance, a wave of migrants came sweeping in from the east via Finland. And, without fail, a number of individuals then end up on the islands.

So let us accept once and for all that Eisen and Stuxberg actually did see a white-backed woodpecker, the first and still the only sighting on Gotska Sandön. To doubt them is envy, pure and simple envy.

Nor can any reasonable person doubt Eisen and Stuxberg's very best discovery, noted in the list without further ado in the section 'Diptera. The larva of a fly'. They succeeded in identifying it as the bee-like hoverfly. '*Microdon apiformis* De Geer. Rare under the bark of deciduous trees.' It is easy to see that a promising future lay before these two young men.

The *Microdon* genus of hoverflies contains species that are both special and difficult to find, and their larvae, which are parasitic on ants under bark or in the ground, look like moving shirt buttons or tiny tortoises. Not even Linnaeus knew what he was looking at when one of his underlings came running up with its larva in his hand. He had no idea. A complete blank. He did not even definitely recognize it as the larva of a fly – or indeed as a larva at all. He fell back on a wild guess and described the beastie as a completely new species, more closely related to slugs than to insects. Perhaps he was having a bad day. Although, to be honest, the larva of *Microdon* really does not resemble anything else.

But the boys on Gotska Sandön knew what they had found all right, and that in itself is amazing. And it reminds me of quite a different story – a story that's still waiting for an ending.

Chapter 6

The Proceedings of the Callicera *Club*

There was something I wanted to understand and then talk about, though not necessarily in that order.

I came very close at times, but just when the answer seemed to be in reach, it slipped away and I watched its flapping coat-tails disappear rapidly round a corner in the distance, so to speak. On other occasions I thought that the question to be understood – and why – was in a sense superficial. The important thing was to recognize a promising trail and to follow it to see what would emerge.

My first attempt arose from the conviction that,

although there is much in this world that is incomprehensible, you can nevertheless discover a meaning as long as you have managed to limit your field of search: by taking up residence on an island surrounded by the sea, for instance, and collecting flies and hoverflies. I am not going to deny a number of rewarding discoveries during that expedition, but I do have to admit that something went awry when loneliness drove me into the arms of René Malaise (1892–1978), the man who gave his name to a fly trap.

I never met him personally, of course, but he was nevertheless my travelling companion, and many was the memorable night he entertained me with the stories of his youth when he was studying Hymenoptera and surviving on pit-roast bear in the wilderness of Kamchatka. The earthquakes crushed everything, but not him. I fell in love a little, in more or less the same way as you can fall in love with your own image in the mirror as long as you don't study it too carefully or too long.

There was no stopping Malaise. His adventures became more and more peculiar, and when he eventually ended up wandering off into the myth of Atlantis, the island that sank, while simultaneously deserting insects and starting to collect old art of doubtful value, I found myself back at square one.

I still own the Rembrandt (according to René) that was stolen from his house on Lidingö during a burglary. It's hanging in my study – a smallish portrait of an elderly man. He greets me with the same sceptical look every morning and then he keeps a watchful eye on all my activities as long as I stay seated at my desk. His repertoire is pretty limited, but at least he doesn't answer me back.

I decided later to return to my old dream of the natural history of a summer's night. If I couldn't find what I was looking for there, then I'd never find it. That I was certain of. But only for a short time. The feeling blew away like one of these fragile truths that flicker before your eyes just as sleeplessness is at last giving way to rest. But the following morning, when you read the half-asleep note you scribbled, with the best will in the world it seems as shallow and empty as a pop song.

A gloomy contemplation of the brevity and vanity of life led to my being sidetracked in the company of the forgotten watercolourist Gunnar Widforss (1879–1934). He was in some respects a pathetic figure and unsuccessful for many years, but kind and honest, generous, too. What more can you ask of a travelling companion? And, in fact, with the passing years he improved as an artist. He died in Arizona, the United States having been his home for many years, and his career as a painter of the national parks was not really so bad in spite of the poverty of the Depression. Or perhaps because of it. They named a mountain in the Grand Canyon after him – Widforss Point. I went there and enjoyed the view for a while.

Gunnar was fleeing from something – that was my distinct impression anyway. From what, though, I never did manage to find out. We got on well together. I wasn't feeling so well myself at the time. I recovered, however, but on the other hand I ended up alone again. Still hunting.

. . .

My collection of flies was now very nearly complete. Growing more and more listless, I spent my summers catching hoverflies I already had. Slowly but surely my trays filled up with long rows that even included the least common species. Since I only collected on the island, that was more or less what was to be expected. Nothing sensational. My big Malaise trap had been blown to pieces in a storm so all I was using now was the net and the pooter, that handy suction device. In the end I simply left the net at home, partly because my technique with the suction tube was, dare I say, perfect. I had learned that success depends on footwork.

The principle is to stand completely still and wait, something I have never found difficult. In addition to which it's essential to stand so that your shadow does not fall across the flowers the flies are visiting, but you must still be near enough so that all you have to do is bend forward incredibly slowly in order to manoeuvre the plastic pipe of the pooter so close to the desired fly that a brief intake of breath causes it to end up in the glass-fibre cylinder. The problem is that it's only too easy to lose your balance, particularly when there is the prospect of catching a rarity, with all the excitement that brings. If you wobble even slightly and are thus forced to step forward in the vegetation, in nine cases out of ten the game will be up. The fly suspects something is afoot and disappears.

But, as I said, I'd mastered all that and the stance I used to adopt was more or less the same as a runner on the starting line of a middle-distance race.

There were still memorable moments. There was, for example, the day in May when the maple was in flower and I managed to suck in a specimen of *Heringia verrucula*, just a little black thing that looks just like many others but turned out to be a first as far as Sweden was concerned. The nearest specimen before that was in Finland, so this boosted my reputation in hoverfly society.

On the whole there were happy endings, but perhaps, I thought, it's time to look around for something different. I certainly wouldn't be the first to give up.

Early in my career as a collector I had acquired five shelf-feet of descriptive lists and related books from an expert on flies who had grown weary of the subject and decided to devote his energies to the translation of Albanian drama and poetry instead. I found his decision incomprehensible at the time, not because I underrated Anton Pashku and the other writers of the western Balkans who would now be available to Swedish readers, but because the entomologist in question was one of the best, one of the real experts, who at the height of his fame went so far as to change his address and rename the house he lived in: he christened it Villa Hottentotta after the eponymous hoverfly.

I have often wondered what the postman and the neighbours thought, but even more often I have wondered what inspired Linnaeus to come up with this strange name. If you are wondering, too, the same family also contains the mysterious species *Villa occulta*, whose presence on the peat mosses of Norrland recently caused something of a stir – a justified stir at that, even if it was among a rather limited circle.

Anyway, it was in that sort of melancholy frame of

mind that I passed my time. The years went by. There was even one occasion when I sat for four days by the stump of a spruce tree out in the middle of a clear felling just to catch an insignificant *Brachyopa testacea* that, embarrassingly, was absent from my collection. I caught it in the end, of course, but the joy I felt was a good deal less than boundless. Something had got lost along the way.

One morning, under a clear sky, the words my grandfather always pronounced in difficult situations suddenly came into my head: 'In the midst of all the bad, the good is being prepared.'

We had been invited out to lunch one Saturday at the beginning of July; that is to say, right in the middle of the season when our island is awash with summer visitors who, with few exceptions, are totally focused on having fun. One party follows another, and in this case the host couple, who own the most beautiful garden on the whole island, are well known for outstanding catering and for their knowledge of fine wines. Their lunches are legendary. As a rule, they begin at the conventional time of one o'clock, but are quite likely to continue until midnight and end with some gentle and rather unsteady dancing to the sound of the bubbling waters of the fountain in the purpose-built summer-house that lies beyond the rose pergola in their paradise garden.

That year's party was no exception.

The result was that I was in pretty miserable shape the following morning. And, as usually happens when I have rounded off the night trying to judge the difference between different years and brands of Calvados, I woke very early, as if driven by an urge to make up for something, though what that something might be remains unclear. The sun was already high in the sky and I decided to go for a walk.

Around eight o'clock I found myself walking past a flowering patch of ground elder where I had enjoyed many happy hours over the years. Just a small patch, about fifty square metres at the edge of the wood. I had caught more rarities there than anywhere else. *Spilomyia*, *Doros*, *Microdon*, *Chrysotoxum*. The grass was still wet from the dew of the night but the sun had reached the flowers so, as usual, I thought I would stay a while. I had my pooter with me. The fly-sucker, as the children called it. My headache was appalling, but it was about to vanish, because, all of a sudden, no more than an arm's length away, there it was. A *Callicera*!

. . .

Of all the world's hoverflies those in the genus *Callicera* are the most beautiful, the rarest and the most mysterious. Big beasts and shy, gleaming metallic-like Byzantine bronze amulets. It's been said many times, but it bears repetition, that even the most persistent collectors need a degree of luck just to see one of them during their lives. That's how arbitrary and erratic sightings can be.

Six species are known in Europe at present, all of them rarities, and at least two of those species (*aurata* and *aenea*) are found in Sweden. Everything about them is rather hazy, but science has come to the conclusion that the larvae develop in damp and rotten holes in ancient trees, most frequently oak or beech, and usually high up. A couple of decades ago, in England, where entomologists have been tireless in their efforts to shed light on the biology of these flies, a *Callicera aurata* larva was found in a small hole a good eighteen metres up a majestic beech tree.

Another British species (*Callicera spinolae*) seems to live in similar circumstances and, since it is thought to have become extinct all over the British Isles apart from in two beech trees in a park outside Cambridge, it has become the object of rather touching care. When one of the beech trees was blown down in a winter storm in 1995, friends of the fly moved in quickly and propped the tree up with stays. A civilized people, the English.

. . .

So there I was, hung over and paralysed. The treasure before me was wandering around unconcernedly in one of the creamy-white flowers of the ground elder. Running on automatic pilot, my brain estimated the distance. It was almost at the limit but it might work. I held my breath and slowly, so slowly, brought the hose of the pooter to my lips.

There is a further strand to this story in that just a few weeks earlier one of my close fly friends had let me know that he was the first person in Skåne to trap a specimen of *Callicera aenea*. He had depicted the event in exhaustive detail and with all the colourful jollity of a Brueghel painting, and I can only assume he had done so for the sole purpose of annoying me. 'It was the late afternoon of a hot June day. At the edge of the woods where I had stationed myself for the time being I could feel the pleasant heat of the sun on my back.' That's how his article opens. He published it in the strangely named journal *FaZett*, that title being an attempt at wordplay, partly on the 'faceted' eyes of insects and partly on the names of two famous entomologists, both long since deceased: Karl Fredrik Fallén (1764–1830) and Johan Wilhelm Zetterstedt (1785–1874).

My friend had taken his stance by a flowering guelder-rose bush outside the town of Hässleholm. His article begins with some marginally relevant chitchat in which he somehow still manages to name-drop a couple of dozen species that were buzzing about among the fragrant flowers of the bush: *Stenocorus meridianus*, *Anaglyptus mysticus* and other beetles, cuckoo wasps, mining bees, daddy-long-legs and, not least, hoverflies, among which were *Xylota abiens* and *Criorhina berberina*. Eloquently, and at some length, he congratulates himself on having found this fantastic bush.

'I was standing there contemplating this and that when suddenly a large dark-coloured fly flew rapidly down from the heavens.' In a flash the author recognizes that he is in the presence of something he has never seen before.

> *The amazingly long antennae immediately caught my attention and when the white tips of the antennae began to gleam in the setting sun my heart jumped a beat with excitement. Stiff and on tenterhooks, I studied the creature. There in front of me, out of reach of course, was one of these fairy-tale creatures I had read about so often and been stung by envy when listening to other people talking about them. Not until now, however, had I ever been granted the pleasure of seeing one in the flesh.* A Callicera.

The article then threatens to go completely overboard before finally managing to save itself and crawl ashore on to solidly scientific ground. The description of the actual moment of capture – a 'leap like a panther' with net raised at the ready – smacks somewhat of over-dramatization, as does 'the Red Indian whoop' of joy that is supposed to have rolled around the neighbourhood a moment later.

There can, however, be no doubt that he trapped the fly. There is a colour picture of it, flanked by a map of its distribution that shows the eight previously known Scandinavian locations. There is even a picture of the actual bush. The literature list is sufficiently long to inspire confidence, and, at the end, along with the expressions of gratitude demanded by the genre, there are some words of acknowledgement directed at the municipality of Hässleholm.

But now it was my turn. If I'd had a net with me the whole business would have been over in no time, but I had left it at home. I would have to rely on the pooter and on experience. All the species in the genus *Callicera* are renowned for their speed. This one could be gone in a flash, that much was obvious, as was the probability that even if I searched for the rest of my life I was highly unlikely to see another one.

And so the pantomime started. All mental activity was dimmed and in a single synchronized movement I leaned infinitely slowly towards the flower at the same time as stretching my arm as far as the hose-pipe would allow. The fly was still there, fifty centimetres to go, forty, thirty, twenty . . .

When the plastic tube was no more than a centimetre or so diagonally behind the fly, I couldn't risk waiting any longer and I sucked for all I was worth. I realized immediately that there was a hitch. Not that the fly escaped – oh, no, not much chance of that, given how hard I had sucked. The problem was that the fly was too big. The diameter of the end of the pipe of my pooter is just six millimetres, which is about as small as you can get away with for the biggest hoverflies, even if you manage to suck them into the tube head first. But if – as on this occasion – they are caught crosswise, what happens is that they stick like a stopper at the end of the pipe and as soon as the suction is released they fly away. I've had that happen to me many times.

So what now commenced was the longest inhalation of my life – I would never have believed the capacity of my lungs to be so great – while my mind worked frantically through all the alternatives. Nothing happened. The fly was stuck across the pipe at such an angle that the air passage was almost completely blocked, which meant that I could carry on inhaling long enough to remember Fritiof Nilsson Piraten's description of a man who was so immensely fat that it looked as if he was always breathing in and never breathing out.

The sun shone.

The swifts screamed.

A jogger ran past along the road.

What eventually happened was that I had no choice but to use my fingers to turn the fly so that it would fit, at which it shot into the tube with a noise like a pea out of a pea-shooter. And at last I could exhale.

I had no difficulty in determining the species.

Callicera aurata, the first specimen in Uppland and the northernmost in Sweden to date.

It would perhaps be an exaggeration to say that my life took a totally new turn thereafter, but everything immediately felt so much simpler. It was not just that my headache had disappeared as I stood there hyper-ventilating in the morning sun, but within a few days I had decided to finish with hoverflies in general and instead focus my studies solely on the genus *Callicera*. And I would take the whole world as my zone of study.

Since the start of the 1980s, when Dr Kumar Ghorpade, a fly specialist in Bangalore in India, described a hitherto unknown species from the western Himalayas, science has recognized seventeen species. They are distributed fairly evenly across the globe, some in America, some in Europe, and rather more in Asia and the islands over that way. Several of these species are still known only from a single specimen. Just one! Not a great deal, then, and what we know about them is on the same sort of level.

The study of *Callicera* is actually the ideal pursuit

for anyone who wants to have a good look around at the same time as contributing to scientific development without having to lug heavy equipment or deal with vast collections of specimens. All that's needed is a net and a pooter, a little potassium cyanide and plenty of time. A certain amount of patience helps, as does a talent for being satisfied with small things or, indeed, with nothing at all.

I also became a member of a society that didn't exist, though that was quite unintentional. Not entirely unexpectedly, rumours of my catch on the island spread with all the speed of the internet and within a few hours I received a message from one of my friends, a doctor in Eskilstuna, one of the exclusive few to have a *Callicera* in his collection. He welcomed me into the *Callicera* Club. That was no more than a phrase he'd come up with on the spur of the moment and it had no reality in the physical world, but the name stuck and later in the autumn of the same year a society of that name was formally inaugurated.

The activity of the club centres very largely around mutual admiration. The sole criterion for election, membership thereafter being for life, is to have caught, photographed or in some other way be able to prove beyond all reasonable doubt that you have encountered a *Callicera* in Sweden. Best of all is to have it impaled on a pin. The club is consequently not a large one. On the other hand the rules say that a meeting of just two members is fully quorate. We normally meet in the Grand Hotel in Lund or, to be more precise, in the restaurant thereof, where we consume good food and fine wines while telling one another what we have caught. This might not look particularly serious but appearances can be deceptive. The idea of the society is that all the younger hoverfly collectors, and I have to say that recent years have seen pleasing developments, will be so envious of those of us who are high caste – if I may put it that way – that they will strive to locate a *Callicera* in order to join us. In the fullness of time that will lead to enhanced levels of knowledge of the biology of the genus. The fundamental and ultimate aim, then, is research.

In the end we may reach a stage where it's possible

for us to locate *Callicera* through the conscious application of knowledge based on research. British experiments with *Callicera rufa* point in that direction. Colleagues over there have been breeding that particular species, found at locations in Scotland, in artificial rot holes.

For the moment, however, it's luck that rules the roost. Of all the qualifications our members possess the greatest gift is to be imbued with good luck. Which is why by far the happiest of all the world's entomological journals is the one we publish: *Proceedings of the Callicera Club*. Quietly boastful, a small circulation, a short and unpretentious résumé in French and a German *Zusammenfassung* – everything just as in the good old days.

Chapter 7

A Session with Charlie Parker

Gustaf Eisen undoubtedly collected numerous hoverflies during his long and eventful life, but as far as I know he never had one named after him. There is a Mexican soldier fly called *Hermetia eiseni*, and in various rather out-of-the-way corners of Central America there are horseflies and crane flies that bear his name, but no hoverflies. But we mustn't get ahead of ourselves.

What came first with him was the constant urge to be best, irrespective of the field, and while still a teenager he settled on earthworms. His inspiration seems to have been Hjalmar Kinberg, a medical man and professor at the Veterinary College in Stockholm, who, like many of those in Eisen's circle, had wide-ranging interests and a penchant for collecting and systematics. He was, for instance, one of the most prominent coin collectors, and among his writings on literature we might note his learned studies into the natural history of the *Edda* and an exhaustive survey of all the domesticated animals that figure in the Talmud.

For a couple of years at the start of the 1850s, Kinberg was the ship's doctor on the frigate *Eugenie* during its circumnavigation of the globe, the Academy of Science having made him responsible for the collection of zoological specimens. For some reason he specialized in birds and worms, and he must have collected a very large number since there were still worms preserved in spirits waiting to be sorted fifteen years after his return home. That became a job for Eisen.

It's certainly a nice tale. Here we have a young fellow, an orphan still at school, a boy seeking the right road, and the road he chooses is the study of earthworms. Or perhaps it was all just chance? In any event, we can be sure that scientists recognized his talent and appreciated his diligence. It is said that he spent all his spare time at the Natural History Museum – which was still on Drottninggatan in those days – where he improved his knowledge of zoology and geology under the tutelage of famous men like Adolf Erik Nordenskiöld, the polar explorer.

He took his matriculation examination in the spring of 1868, the dissertation on Gotska Sandön being published at about the same time, and in September he enrolled at Uppsala University.

The zoologist Vilhelm Liljeborg now took over as his mentor, along with the botanist Thore Magnus Fries, and it occurs to me that scientists are reminiscent of jazz musicians in the sense that names are so important. Idols and pioneers who smooth the road. Once you've played with Charlie Parker you tell everyone about it. And it's quite possible that in Eisen's day John Areschoug, for instance, was a sort of Dizzy Gillespie of brown-algae research, or that in his own sphere the arachnologist Tamerlan Thorell was as much a risk-taker as the saxophonist Lasse Gullin. How should I know?

There are, however, three names that simply cannot be ignored as we set out into this world, three legends of natural history – a Swede, an Englishman and an American – all born around the beginning of the century and thus about forty years older than Eisen: Sven Lovén, Charles Darwin and Louis Agassiz. That's what I call a trio.

. . .

After three years in Uppsala, Eisen was ready to publish his first major work on earthworms, *A Contribution to the Oligochaeta Fauna of Scandinavia* (1871). An impressive piece of work. All of the species known at that time are covered in a manner that bears witness to the author's familiarity with the sources from Linnaeus on and especially to his own experience of worms both in the field and under the microscope. He describes everything there is to see and to know in Latin and in Swedish. Of particular interest for our present purpose is what he has to say in the sections entitled 'Habitat and Way of Life' and 'Distribution': it is possible to garner from these a good deal about

Eisen's own habitats as well as some aspects of his way of life.

It emerges, for instance, that he was now twenty-four years old and had managed to travel far and wide around Sweden and Norway; also that he was corresponding with eminent worm scholars on the Continent. Detailed descriptions of the often nocturnal activities of the various species tell us a bit more about his habits and remind us once again of Strindberg's recluse. It becomes possible for us to work out exactly which species he was after on the nights his friend Strindberg was allowed to accompany him.

We can imagine the yellow gleam of their lantern in the summer night and hear their whispering voices. Eisen writes:

> L. foetidus *seems to be quite uncommon here. I have only found it in a few places, but then in great numbers. It would seem to be most common in Skåne and Blekinge. I have also found it at Alingsås and Kinnekulle in the province of Västergötland; also in the botanical garden in Uppsala, where it is present in very large numbers.*

L. foetidus *is one of the rarer species in Germany, France and England.*

Lumbricus foetidus. I read the name aloud to myself, and the thought suddenly struck me that now is as good a time as any to tell the story of our house on the island. Lovén, Darwin and Agassiz will have to be patient for a moment, because I actually have a history of my own that involves this nowadays very common species, also known as the brandling.

Apart from in this particular case I have kept my relationship with the biology of earthworms firmly under control. I have never been a fanatic and I'm unlikely to become one. It would never occur to me to collect earthworms: they must be quite problematic to handle, and the currently popular alternative, which is to photograph things – butterflies or whatever – rather than collect them, doesn't really attract me. Truth to tell, I can walk past worms without a second glance. In so far as I feel anything at all it is a sense of how tedious they are.

Before we moved to the island the place of earthworms in my world was wrapped in vaguely romantic memories of past summer holidays and sunny days when successful fishing was wholly dependent on access to worms for bait. Waders and bait boxes.

The smell of earth and stinging nettles. The tarred wood of the landing stage in the warmth of the sun.

We did a lot of fishing, my friends and I. It took no more than ten minutes to cycle to the Gränsö Canal, which was still idyllic in those days, with its steep overgrown banks, particularly south of the old wooden and iron bridge that had to be opened manually. That was where I saw my first kingfisher. It's an old canal, dug by captured French pirates. That's where we fished. Perch mostly, and roach. So I knew something about the habits of earthworms.

But now it was 1986.

I was still working backstage at the Royal Dramatic Theatre in Stockholm and living in a sublet flat not much bigger than a clothes closet in the Kungsholmen part of the city. Meanwhile, Johanna had moved up from Skåne. In love we might have been, but since we were expecting our first child it was high time to start looking round for something bigger. Buying a flat anywhere near the centre was out of the question, given that our income was far from generous, but we thought we might just be able to find a cheap house somewhere out on the edge of the city.

There are two ways of going about these things. You can either spread the search over a large area and have many places to choose from, or over a smaller area where the choice is limited. To avoid running out of energy we decided on the latter. Few activities are as wearisome and stressful as looking for a house in which you might spend the rest of your life. You may not be given to vacillation at the start, but you soon will be.

I'd been out to this island earlier, though only for short visits. The first time was in May 1984, when I went there to write a newspaper article on the unique orchid flora for *Stockholms-Tidningen*, but, since I was already in half-hearted flight from biology and science by that point, the article, if I remember rightly, ended up being mainly about Strindberg and the cottage in which he spent a summer writing, swearing and yearning. That's where his novel *By the Open Sea* was composed. Other things, too. He wrote the strange novella *The Silver Lake* on the island, as well as many letters. I ended up sitting in the sun, fantasizing about the future. There was something special about the island.

Which is why, two years later, we decided to look for a house on the island and nowhere else. That made things easier, because there were so few things on offer. When we started in April 1986, there was only one house for sale and it wasn't a place we wanted. We went there anyway and wandered around talking to the people we met and asking for advice. We are thinking of living here, we said, and it must have been pretty clear to anyone and everyone that we needed somewhere before the autumn. In the end a woman resident told us about a house that hadn't been occupied for thirty years. Pretty dilapidated, of course, but still . . .

We walked past it. It was 6 May. The cowslips were in flower, the cuckoo was calling, and the holly-blue butterfly had just taken to the wing. The house proved to be completely ramshackle, but the plot it stood on was paradise.

We discovered from the property register that the house was owned by a middle-aged man living on Gotland. We wrote a long letter. He answered – the tone was rather correct but he also sounded delighted – and he told us the history of the place all the way back to when it was built at the turn of the century. While still a child he had inherited it from his maternal grandfather. Now it was empty and dilapidated and the garden that had once been so beautiful lay in a long deep slumber overgrown by alder scrub and shaded by enormous spruce trees. Trees had taken over and it was hardly possible to see the lake any longer.

This friendly letter from the owner naturally raised our hopes, and, although he avoided directly answering our enquiries about the possibility of purchase, the correspondence continued throughout the early summer, until, one beautiful August day, he suddenly let us buy the house. The money itself had not been the important thing; he had just needed to get used to the idea. We got the place cheap.

To our amazement this man became famous, or notorious rather, about ten years later. In his dealings with us he had very obviously been driven by passions other than economic gain, but in a different context he had been afflicted by that special form of lunacy that reveals itself in the theft of rare books. He was a man of considerable culture and highly respected in Visby, where he lived in a stone house from the thirteenth century. The oldest house in the town, he used to say. But then he had started stealing books, one of them being Isaac Newton's *Principia*, a first edition from 1687 that had been donated to the Visby Public Grammar School.

Disaster struck later, when he sold the book for £60,000 at Sotheby's in London. Everything came out and our benefactor ended up in prison, latterly also spending periods of time in mental hospitals. He is no longer alive, but he used to ring us occasionally once he was out of custody and he was always cheerful and friendly.

Anyway, we immediately moved out to the Stockholm Archipelago and made the island our own. Before we built the house by the lake we lived in the cottage for many years. We couldn't afford to do anything else. The roof leaked, the insulation in the walls consisted of old ants' nests, the well was little more than a lethal hole in the ground, and since the first winter was the coldest for 170 years we had to take an axe in order to fill up the buckets. No one felt any great desire to sit and philosophize in the privy behind the woodshed. There was electricity in the house, but that was the lot.

The privy, now, with a growing family. Handling the privy containers quickly became heavy work, although rather soothing in a way. I mixed the contents of the containers with leaves, sawdust, grass and all the other waste produced by a country garden, and the earthworms of the district converted the compost heaps into the very best topsoil you can imagine. By the early 1990s the dungheap was the size of a Volkswagen, or maybe just a Fiat 500, and I estimated it to be quite big enough to hide at least three fully armed security men from the Secret Service.

. . .

Because at that time I was keeping company with Al Gore, at a distance anyway. He had just become Vice President of the United States and brought out his book *Earth in the Balance* (1992), which dealt with solutions to the world's environmental problems. It wasn't that I agreed with everything he had written, but there was something about the tone of his book that appealed to me, something about its passion. We were in need of money, so I took the book to Bonnier's and offered my services as translator. Agreed and done, a real doorstopper of a book. Then, before the launch of the book, there was a rumour that Al and his wife, Tipper, were going to visit Sweden and there was talk that we should show them something of our beautiful country while they were here.

Put them in a helicopter, I said, and take them for a spin over Stockholm and the islands of the archipelago. There's nowhere more beautiful. And if they land by our lake the author will be welcome to a cup of coffee and a chat with his translator in the latter's humble home. They thought this was a brilliant idea. Our lake isn't exactly Walden Pond, but it's not far off. The publisher sounded out the Foreign Ministry and everyone seemed to be singing off the same hymn-sheet.

But it fell through. They didn't come, not even to Sweden.

The only thing that really vexed me was the business of the bodyguards behind the compost heap. I had become so fixated on this fantasy that I couldn't walk past the compost for ages afterwards without looking to see whether anyone equipped with telescopic sights and a bullet-proof vest was hiding there. There wasn't, of course.

. . .

The translation, however, provided the first slice of funding for our new house. The time was at last ripe for a modicum of material comfort, with running water and plenty of room for children and books. The days of the privy in the garden were definitely over. But since we had now got used to having virtually unlimited quantities of compost, instead of going for an ordinary three-chamber septic tank, we acquired an advanced composting system and installed it in the cellar, directly beneath the bathroom. I won't go into detail but it consists essentially of a large plastic container in which an army of earthworms was supposed to manufacture topsoil at the same rate at which we provided raw materials from the floor above.

I learned two things at this stage. In the first place I discovered that the breeding of worms has become an industry. *Eisenia foetida* (the genus *Lumbricus* changed its name to *Eisenia* early on), the compost worm known worldwide, is bred on worm farms, from which it's possible to order suitable quantities by post. According to the advertisement, all you need to do next is to introduce the animals into the dung and all your troubles will vanish.

The second thing I learned was that this rarely if ever happens. You have to spend the whole time fiddling and pottering with the worms; otherwise they die.

In line with my usual practice I tried reading around in the hope of acquiring the right touch with these pets, but the books I read merely reinforced my suspicion that modern composting is actually some sort of philosophy of life and consequently not something for me. We now have a three-chambered septic tank. Walter Buch, a German expert, wrote quite dispassionately that in the old days dried earthworms were ground down to a powder that was then mixed with gunpowder in the charges used in firearms and cannon. This was believed to increase accuracy.

No doubt it worked as long as you believed in it. And, it may be unjust of me, but I have a feeling that more or less the same thing is true of earthworms in latrine systems.

. . .

Gustaf Eisen was a Darwinist and one of the first people in Sweden to really understand the import of the theory. That is probably why he mustered the courage to send his thesis on Scandinavian earthworms to the master in Down House in Kent. His accompanying letter has long since disappeared, but Darwin's thank-you letter to him still exists. It is dated 3 December 1871. A relic. A fetish. Indeed, this letter echoes like a distant saxophone solo in the background through all the seventy years that Eisen had left to live.

This letter came up in conversation with the journalist Erik Wästberg when he visited Eisen on Park Avenue at the end of the 1930s. The old man in the enormous apartment had just told him how he had once met the king – Karl XV – and then he proudly showed Wästberg a letter from his friend Strindberg in which the latter writes, 'Gustaf! What you have done for me you can be bloody sure you haven't done for nothing . . .'

Then Eisen produced another letter and began talking about Charles Darwin.

Just before the war Wästberg wrote an article about Eisen, drawing together all the threads, and there can be no doubt that the letter from Darwin was like a session with Charlie Parker. The old man was still proud of it, even though it wasn't much more than a short acknowledgement. Darwin had been impressed, both by the colour plates that Eisen had drawn himself and by the detailed information about the distribution and behaviour of the animals, which Darwin had had translated into English. I, too, do some work with earthworms, Darwin wrote.

Chapter 8

The Natural History of the Summer Night

The woodwork teacher known as Woodwork Hitler gave a brief smile and considered my request. The issue of the day, as usual at the start of term, was what plans the pupils might have, what did they want to do in woodwork. The choice was up to us, within reason of course, and my classmates respectfully came up with a whole series of ideas about cutting boards, fruit bowls and dish racks. Woodwork Hitler nodded approval in a disinterested manner and with no change of expression directed them to the dog-eared folders that just happened to contain ready-made plans for cutting boards, fruit bowls and dish racks. Not until it came to my turn did he reveal a crack in his façade.

Planed lime-wood, ten centimetres wide, and, if that wasn't available, aspen. The wood had to be soft, I said, doing my best to sound firm. His thoughts wandered, no further than the wood-store in the next room perhaps, but in retrospect I have a feeling that he suddenly remembered something long forgotten: what it was like to be a boy with something in mind.

I had my copy of Carl H. Lindroth's informative *Guide to Collecting Insects* with me, and it helped to explain why the variety of wood used for butterfly mounting boards needed to be a softwood like lime. Excellent, he understood that. He had lime-wood in the store and, if I'm not mistaken, there was a tone of approval in his voice when he gave me a sheet of squared paper and told me to make a detailed drawing. There's not much oomph about a cutting board!

I had made mounting boards before, so I'd finished them in the first two lessons. Now was the time for my secret plan. My lamp was going to be ready long before the butterflies began their summer flights, but

I'd have to be cunning if I were to have any hope of success. To come straight out with it and ask the teacher's permission to make climbing irons out of steel reinforcing bars would be pointless, because, even though a certain amount of metalwork was part of the course plan, it didn't take much imagination to see that the answer would be a straight refusal. Or a refusal after first interrogating me about what I wanted them for.

Electricity Board engineers had climbing irons, as did the telephone company people, but they were strictly forbidden to ordinary people. They could be lethal. Not so much the climbing side of things, perhaps, since the design is quite straightforward, being just a sort of metal hoop that attaches to the sole of your shoe. No, the real danger lies in the high-voltage cables you can climb up to using that sort of equipment.

The guidelines I had drawn up for myself in the loneliness of the night involved avoiding making a frontal attack on my target. For tactical reasons I had decided to make an indirect approach. The idea was that I would first construct a large number of nest-boxes. No one could object to that. With these ready in time for spring – big and small, starling boxes, tit boxes, treecreeper boxes and a few specially designed for owls and stock doves – I was intending to legitimize my deception with the help of another book I had read very thoroughly.

It was called *The Nest-box Builder* and was written by Torgny Swiss Östgren. In the section of the book that deals with how high above ground different boxes should be positioned, it states that small birds are satisfied with nesting locations that can be reached with an ordinary ladder, whereas stock doves and tawny owls prefer their residences to be fifteen metres up the tree. How do you get up that far? The author didn't say, but I had an idea. I was finding it impossible at that stage to come up with an absolutely watertight plan, but the very fact that my strategy was a long-term one, taking something like six months, convinced me that it would work.

The late winter of that year found me making nest-boxes as if there were no tomorrow. It was fun and it wasn't difficult. I would go so far as to say that some of the models exhibited a degree of elegance. Unless my memory is deceiving me, that was particularly true of my copies of the so-called Homely Box that had just been launched by the Gävle Workshop for the Disabled as a sort of great tit equivalent to the almost hysterical house-building programme (the Million Programme!) that was going on in the suburbs of our towns at the same time.

I saved the owl and stock dove boxes until last, and when they were ready I approached the woodwork teacher as planned and asked permission to complete the project by making climbing irons. Carefully and eloquently and with much reference to the relevant literature, I drew his attention to the dire situation of the stock dove in our forests, where the competition for holes fifteen metres above the ground was fierce.

His answer was no. No further discussion. End of.

And that was that.

I'm forever getting myself entangled in complicated strategies that sometimes work but that remarkably often fail and end up leading me along unforeseen paths. I never did get hold of a really good moth lamp, not as a child anyway, but it wasn't that important. My intentions were secret and remained so: as far as others were concerned I was just a happy nest-box builder. The result of this, however, was that I began to take more of an interest in birds and very soon developed considerable expertise about the rich fauna of beetles, fleas and other insects that live exclusively in birds' nests.

I would actually go so far as to concede that my inability to reach that lamp was ultimately more rewarding than if I had been successful. Not with regard to moths, perhaps, but in other ways.

What the whole business then developed into was something of a fixation with anything and everything that moved under street-lighting. I learned to distinguish different kinds of lamps at a distance and even now I believe I can tell from an aeroplane at many thousands of metres at night whether the lights along a road are neon lights or mixed-light lamps of the kind that go out when you kick the lamp-post. It's a good trick, especially when there are girls to be impressed, and we used to gather in the evenings under a street-lamp closer to Gränsövägen and devote the night to scrumping fruit.

And every time I am sitting in the window seat of a jet at night as it passes high above a town bathed in a flood of light, I remember my last and greatest triumph before leaving childhood and Västervik and setting out into the world.

. . .

We were all beginning to grow up. Hardly anyone did things on their own any longer. Life in the group I kept company with was characterized by a kind of capercailzie display. All the moths and butterflies in the world couldn't help me any longer, but I still retained my interest in street-lighting. Kicking a lamp-post in town, where we spent our nights hanging around, and so extinguishing a light here and there along a cycle path no longer impressed anyone. There were so many lamp-posts everywhere that it was impossible to summon total darkness, and the girls wouldn't just let you kiss them for nothing at all.

Finally, however, I discovered how to go right to the top level.

I'd begun to apply some theoretical thinking to it all and wondering, for instance, how all these lamps were switched on at night. The idea that there was someone at the highways department who got out of his armchair at the right moment and flicked a switch seemed to me highly improbable. I can't remember exactly how I discovered the answer to the question, but somehow I managed to access the necessary information, which was that the lighting for the whole town was controlled by a photocell, a sort of automatic eye that was sensitive to the fading light of dusk and the growing light of dawn.

Västervik is not a big town so I began searching for it. I had no idea at all what a photocell might look like, but I thought that it probably wouldn't resemble anything else and would be positioned in the open, on a post, perhaps, or the roof of a house. Later, I heard a rumour that the photocell was located somewhere near water, possibly at the harbour or down by the bay at Gamlebyviken. That's where I concentrated my search.

And one fine day I found it, on the north-facing wall of a house more or less opposite the baths and straight across the water from the island of Strömsholmen. It was a less-than impressive object, fixed quite high up but low enough for it to be easy for me inspect how it worked. I waited until dark and then shone my pocket torch on the photocell: the whole town suddenly blacked out.

Oh, what a wonderful feeling!

Ambitious I might be, but I've never been hungry for power. Ever since that evening, however, I know quite a bit about the sweetness that is said to accompany the sense of absolute power.

Electronics has no doubt become much more sophisticated since, but at the time I was able to extinguish the whole of Västervik – it was the summer of 1976 – the photocell simply assumed that the light from my torch meant that morning had suddenly broken and it was time to switch off the lights. And when I then aimed the beam of my torch in a different direction the silly cell must have assumed that it had been a bloody short day, because it only took a few seconds for everything to light up again in all its glory.

I tested it once more.

It was just as much fun the second time.

This story has a sequel that doesn't really belong here, but since I know that the event later became the subject of a good deal of speculation even on the part of the police, a few words might be in order. For information only, nothing else.

It was a warm evening in late summer. There were still many sailing boats at the visitors' moorings, and the wooden quay on Strandvägen in front of the grand Häggblad House had its share of leisurely strollers because there was a touch of Indian summer in the air. It was the sort of evening that might be the last of its kind before autumn arrives and the serious time of year begins.

We took great care with our preparations. A friend and I had stolen a mirror from a public lavatory down on the square and we rigged it up so that it reflected the light from a street-lamp straight into the eye of the photocell. From then on everything went with the precision of a pianola. The moment the lights came on the photocell assumed it was morning and switched them off, and within a few seconds of their going off they were switched back on again. This sent the photocell into a blind panic, and morning and evening began to follow one another in unnaturally rapid succession. Our lovely town started flashing on and off like a pinball game or Las Vegas Boulevard late at night.

People out for a walk along the promenade stopped in amazement, but the girls smiled on my friend and me.

. . .

Years later, after I'd travelled the whole world and the time had come to decide on an academic specialization, I was overwhelmed by grave doubts when it came to setting my sights on a distant goal along a road that would lead to examinations and titles and, it has to be said, self-respect. Until then my study of biology had been quite unfocused.

The new friends I had made at university in Lund all seemed to have obvious careers in mind. Many of them were thinking of becoming teachers, others officials in the Nature Conservancy. Right from the first term one of them was determined to cultivate pelargoniums and other pot plants, while several others were already moving in the direction of the pharmaceutical industry. There were also many who wanted to be research scientists, and some of them succeeded. You catch an occasional glimpse of them still ploughing the same furrow. Gaunt professors, mostly ecologists, since that was the fashion of the day. The academic world is as tight knit as a colony of puffins.

I, too, was attracted by research: systematic zoology in particular, especially insects. Just then I was back in one of my recurrent beetle periods, but the more I saw of the Department of Zoology, where the stuffed Great Auk seemed to set the tone among the staff, the more hesitant I became.

It came to a head one morning in May with my buying a bottle of the cheapest red wine and cycling out to the nature reserve at Fågelsångsdalen, on the way to the lake at Krankesjön and the Vombs Ängar water meadows. Once there, I lay on the slope on the north side of the burn where the grey wagtail nests and drank the whole bottle, pondering deeply all the while. What was it I really wanted? That was when I recognized that my subject was the natural history of the summer night. But there was no such subject, so I stopped.

Chapter 9

Lucky Boy

Of all Darwin's books the one I rate highest is *The Formation of Vegetable Mould, through the Action of Worms, with Observations on Their Habits*. Published in 1881, it was the last thing he wrote and it's a classic among biological writings. His 1859 *Origin of Species* was, of course, epoch-making, one of the most influential books of all time and as enduring as stone, but *Worms* is more beautiful. As someone once said about Thoreau, it's clever and it's hopeful.

Here we have a man who has achieved everything. While he was still young he made a long journey and saw the world and now the world has seen him.

Powerful men of the Church are still resisting and publicly defaming his work, but he knows who will win in the end. He is less and less concerned about the disputes, tired of all the squabbling, and no one, least of all him, could bear the weight of more honour on his shoulders. His stomach is playing up, as usual, and so he retreats, potters about in the garden and writes the book about worms he has been intending to write for at least half his life. At last he is a field biologist again.

I think the key to the book is to be found in a letter. Well, there are probably many keys, or perhaps none at all, but my reading of it was illuminated by the advice Darwin gave his son George when the latter, with all the impatience of youth, wrote a dissertation critical of religion and asked his famous father to cast a critical eye over it before publication.

Stop and think, Darwin wrote in reply. Lay the pamphlet aside and wait a while. Criticizing the Church by attacking it directly will be of little lasting value. It's like pouring water on a goose. Not even Voltaire succeeded and his pen was razor-sharp. Or, he continued, pointing to both John Stuart Mill and Charles Lyell, you might just as well simply desist, not bother at all. But at the very least, he recommended, limit yourself to 'slow & silent side attacks'. Underground activity. Do what worms do. They create land from the bottom up, literally, silently and persistently, at night.

The wonderful thing is that the book can be read as the purest of pure science, composed for no other purpose than the search for knowledge and to pass the time when the time is limited. Patient studies of the worms that are munching away at the lawn seem to me to be particularly appropriate for a man of his age and disposition. He walks the gravel paths like an ancient guru, his mind pondering insignificant things.

What do we know about the intelligence of worms? He investigates the topic. It takes time because he is old, but he eventually proves that they are more intelligent than one might expect. But they are also stone deaf – he manages to discover this by using a famous series of experiments. I can't help visualizing the process as a short film and his account of his results should be inscribed in letters of gold over the door of every research laboratory:

> *Worms do not possess any sense of hearing. They took not the least notice of the shrill notes from a metal whistle, which was repeatedly sounded near them; nor did they of the deepest and loudest tones of a bassoon. They were indifferent to shouts, if care was taken that the breath did not strike them. When placed on a table close to the keys of a piano, which was played as loudly as possible, they remained perfectly quiet.*

One can only wonder what piece was played. It must have been Darwin's wife, Emma, who sat there hammering at the piano so that the whole house shook. He could presumably manage the whistles and shouts himself, probably some of the oompahs on the bassoon, too. But he was more or less tone deaf and lacked all sense of rhythm, so he usually left the piano to Emma, who had taken lessons from Frédéric Chopin in Paris when she was young.

Eisen's dissertation on Scandinavian earthworms

now came in handy. Darwin makes an early reference to it in the first chapter of his book and a number of times thereafter, which must have been very flattering for Eisen, as well as being a welcome acknowledgement that carried more formal and scientific weight than the letter he had received in Uppsala ten years earlier. And by this time he really did need some encouragement. Eisen was out of luck: not only did he lose his parents but he was soon to lose his fortune too. And, somewhat later, almost everything else.

But let's return for a moment to his happy student days in Uppsala at the start of the 1870s, to a time when scholars with power in academia were planning his future and everything seemed to be falling into place for him.

One of the professors who had most to do with his success was Sven Lovén, curator of the invertebrate section of the National Museum of Natural History and an internationally renowned expert on sea cucumbers and the like. He had been elected member of the Academy of Science back in 1840 and made a professor the following year while still in his thirties. It seems pretty clear that he saw the young Gustaf Eisen as his potential successor. A man of the future. Lovén was the sort of man who spurs others on, supports them and despatches them on long voyages. While Eisen was still in the sixth form and frequently hanging around the national museum he happened to hear Lovén say to Nordenskiöld: 'That boy has remarkably sharp eyes. He'll be a naturalist for sure.'

After he'd been in Uppsala for five years no one was left in any doubt. Eisen had shown his talent both as a collector and as a systematics man. He had started describing new species according to all the rules of the game, which demands a good deal of experience and, above all, an overview – knowing what everyone else has written about the particular group of organisms one is working on.

He christened his first species *Enchytraeus ratzeli*, naming it after Friedrich Ratzel, a colleague on the Continent and a dynamic scholar who would later be spoken of as the founder of cultural geography. That same summer Lovén came up with the funding to send Eisen to England, where he worked for a time under

John Edward Gray, the legendary director of the zoological collections at the British Museum. Gray was also known as an early philatelist; very early in fact: when the Penny Black, the world's first stamp, was issued in 1840 Gray bought several of them on the day of issue.

Unlike Stuxberg, Strindberg and the other hard drinkers in Uppsala, Eisen took his degree in good order in May 1873 and was made reader in zoology virtually simultaneously. He was twenty-five years old and stood on the threshold of an adventure. A research trip to Boston and from there to California. Two years, that was all, then he could take over. His course was marked out for him.

. . .

Strange things turn up in the newspapers. There are days when next to nothing happens, but the columns still have to be filled with something or anything – and that is more or less what happened the first time I was written about. Pure chance. The article was in the *Västerviks-Demokraten* one morning in July 1968. 'Lucky

Boy Catches Two Pike with a Borrowed Rod' was the headline. There's also a picture of me with the two pike. Being a journalist must be a nightmare sometimes.

It's obvious, though, that I was lucky. I still am.

We'd been out in an old fishing cutter called *Ringskär*. During the Slottsholmen Song Festival the boat was used to transport intoxicated troubadours out to an island in the archipelago – I don't recall which island – where they would quickly get even more drunk and sing to the high heavens before sailing back to the town in the small hours. For some reason or other we – the whole family, that is – were on one of these trips and the local press was covering it.

Once we were out on the island I ran around for a while on the rocks that lined the shore and then later, as the sun was setting, the skipper of the boat lent me a fishing rod. That was kind of him. I'd never held a casting rod before so he showed me what to do. I shan't forget the scene. The sun had just set; a landing stage by a boat-house in a bay where the greenish-black water was like a mirror; the heat of the day was still lingering and with it the smell of tar, seaweed and mud. I caught a pike with my first cast, quite a big one, and another smaller one just a short time later. People congratulated my parents.

'The boy's lucky.'

I acquired a rod of my own a couple of years later, but the miracle was clearly over, temporarily at least, because I didn't catch many fish to get excited about. Not enough for the papers to think it worth writing about, anyway. Instead, I learned all about my hook getting snagged on the bottom. Every time my nylon line broke and I lost my lure something inside me snapped, too, which is why I hardly ever used anything but small light spinners or tiny spoons that didn't sink so quickly, although they had the disadvantage that you couldn't cast them more than a couple of metres.

Only once did I possess a larger lure – a wobbler, the name alone is an adventure – and at the very first cast there was something miraculous about it. Because of its weight and its cigar shape, it flew like a bird for fifty metres, perhaps even more, out into the bay at Grantorpsviken. The wobbler curved out over the water in a wide arc, providing me with a moment of joy and strength. I'd never managed a cast anything like it before.

Only when I began winding in did I realize that the line had twisted and snapped at the very moment of casting. I've come across this phenomenon later both in illustrations and in literary sketches: the best casts are sufficient unto themselves in the sense that they are invisible to the man holding the rod and only the onlooker can fully appreciate them.

That is the kind of arc followed by Charles Darwin's autobiography in my library. I love the book, *Recollections of the Development of My Mind and Character*. He wasn't actually trying to catch anything, just to give an account of himself to his nearest and dearest. A medley of anecdotes. Nothing out of the ordinary. There was no question of publication, not while he was alive.

Chapter 10

The Third Island

The year was 1873. Charles Darwin had withdrawn to Down House to nurse his troublesome stomach and to write about the habits of earthworms in peace and quiet. Gustaf Eisen, meanwhile, was twenty-five years old and on his way out through the door and off. He wouldn't see his native country again until 1904, a little over three decades later.

Once again it was Sven Lovén at the National Museum who lay behind his travels. Eisen, of course, had two half-brothers in San Francisco and he could also afford to pay a good proportion of his travel costs with what he had inherited from his father, but the organizer of the expedition was undoubtedly Lovén, the man who extracted funding from the Academy of Science and wrote the necessary letters of recommendation.

As part of the agreement, Eisen was commissioned to collect molluscs and other specimens on the West Coast of America for Lovén. A number of professors in Uppsala contributed smallish sums in exchange for a promise to collect various other things: algae for Areschoug, lichens for Fries and spiders for Thorell. In a letter home to his friend Stuxberg, Eisen reports that he had been out to shoot an antelope for 'old Liljeborg'. It is Eisen's letters to Anton Stuxberg that provide us with the most reliable picture of what he was up to in America during the 1870s.

Initially, anyway, everything went according to plan. Autumn found Eisen at Harvard, near Boston, where he was welcomed with open arms by no less a figure than Louis Agassiz. That, too, was set up by Lovén. It would hardly have been possible to have a better start.

Agassiz was a living legend and the leading scientist

in the New World. He was born in Switzerland in 1807, studied medicine in Germany, but at the same time he kept up his zoological and botanical interests. While still a young man he moved to Paris, where his talents were noticed by the palaeontologist Georges Cuvier and by that all-round genius Alexander von Humboldt, both of whom were born in 1769. Under their influence Agassiz was attracted to geology and the question of how life had developed. Unlike Darwin he was never able to free himself from the religious preconceptions that stood in the way of a tenable theory, but despite that he was the man who discovered that the Northern Hemisphere had been much colder and widely glaciated during an earlier geological period. Agassiz was a professor in Neuchâtel for many years but he went to the United States on a lecture tour in 1846 and remained there, shortly afterwards becoming a professor at Harvard.

By this point he was at the height of his career. He had founded the Harvard Museum of Comparative Zoology and was one of the founders of the American Academy of Science. There was no shortage of money, but talent was in short supply. Agassiz himself may have been a brilliant scientist but, on the whole, America still had little to contribute. 'Natural History here is completely based on humbug,' is how Eisen describes it in a letter. Education was in a very poor state, and anyone hoping to build strong universities was forced to employ rising young men from other parts of the world.

Enter Eisen. Agassiz was delighted and impressed by the young Swede, I think. It may well be that he recalled the support and encouragement he himself had received from Humboldt and Cuvier, and thought it was time to repay the debt. Within a matter of weeks Lovén must have felt that he'd shot himself in the foot by sending his intended successor to Harvard because Agassiz had already recognized what a find Eisen was. Since Eisen was going to the West Coast anyway, Agassiz offered him more money to collect specimens for Harvard than he had received from Sweden. Half the sea cucumbers would go to Stockholm and half to Harvard was the new arrangement.

Eisen was given a sizeable advance, and Agassiz also

provided him with equipment such as jars, bag-nets, spirits and anything else he might need in the course of a two-year expedition. The whole lot was shipped round Cape Horn to San Francisco. Agassiz certainly baited his hook, there's no doubt about that, and when, shortly before Eisen's departure, Agassiz offered him a professorial post at Harvard on his return Eisen was ready to burst with pride.

There was something very special about this hard-working son of a wholesale merchant. Everyone wanted him. Full of confidence and enthusiasm, he took the train west. He crossed the Missouri at Omaha and was amazed by the mounds of buffalo skeletons whitening on the prairie. Then it was over the Rocky Mountains, through the deserts and the Sierra Nevada, and down to San Francisco. By now it was Agassiz who was writing his letters of recommendation, and they opened all doors.

Among the first things to happen in San Francisco, almost before he'd had time to say hello, is that Eisen was elected a member of the California Academy of Sciences. It's important not to forget how long ago this was: Billy the Kid was still in short trousers and San Francisco, which before the gold-rush of 1848 wasn't much more than a collection of tents and shanties, still hadn't begun to build its new avenues and substantial stone buildings. This was a land of settlers and pioneers. Money was what counted. The town guard patrolled the streets at night and the bandits they caught were hanged at dawn. Earthworm research was not well developed.

Then, just before Christmas, Louis Agassiz died quite unexpectedly.

This was a major blow for Eisen, but at least he could fall back on his original plan. He began to collect specimens close to the town and then rode further inland together with his half-brother Francis, a businessman who had just bought a large area of land at a place called Fresno in the San Joaquin Valley, within sight of the snow-capped peaks of the Sierra Nevada.

But it was marine fauna Eisen had come to study, and at the beginning of March he moved south along the coast and set up his base, astonishingly enough, on Santa Catalina, the small island off Los Angeles.

Nothing astonishing about it for Eisen, but it certainly was for me, because that's where the watercolourist Gunnar Widforss settled at the start of the 1920s. He travelled to San Francisco by train and then proceeded straight to Catalina, by then a flourishing tourist trap owned by the chewing-gum magnate William Wrigley. A mysterious coincidence was my first thought, but then I quickly remembered the significance of simply sitting still and waiting, for stories or for whatever. Sooner or later everything seems to be part of the same puzzle.

In Eisen's day, however, Santa Catalina was still uninhabited. Pure wilderness, just as unexplored as Gotska Sandön. Since the owner of the island had taken it into his head to donate the whole island to the California Academy (various intrigues prevented this actually coming to pass), it seemed appropriate that an intrepid naturalist should be sent to inspect the place. That suited Eisen down to the ground – an island is always an island.

Spirits and bag-nets. What more can a man ask for? Before the summer was over Professor Lovén received more than he could have dreamed of. Sea urchins alone filled a barrel, and Areschoug got his brown algae, the largest of which – a sort of kelp – was new to science: a few years later it was given the name *Eisenia arborea*.

Few people are privileged to have one taxonomic family named after them; to have two with identical names, *Eisenia* (worms) and *Eisenia* (algae), is unique. All in all Eisen gave his name to five families, one subfamily and a very large number of species.

He was on his way: 'An abundance of specimens of all kinds of things.'

Letters from America constitute a genre all of their own. Their writers are always keen to present themselves as strong and successful. But it wasn't all peace and joy on the island. Eisen wandered around the shores like some latter-day Crusoe, but as time went on he was plagued by loneliness and homesickness. In spite of everything he wasn't perhaps the ideal collector, not made of the same stuff as those hardy Brits – Wallace, Bates and the rest of them – able to endure years in the wilderness with nothing but their butterfly traps and bird guns.

Nothing seems to have engaged Eisen's attention more during this time than Stuxberg's drinking habits back in Uppsala. His letters are a mix of fatherly advice, cajoling and financial inducement. He is quite obviously still acting as his friend's private financier since he is remitting a hundred riksdaler a month via one of his half-brothers. The only thing he asks for in exchange is the occasional letter. He also asks for a photograph. I'm glad the letters exist: they tell us, for instance, what happened when Eisen was ruined, when he lost everything overnight.

But first he had to go to the mountains, and that, too, was an adventure.

Tired of the dry yellow hills of San Francisco, of the fog and the summer wind that whips up sand and dirt, I decided in short order to pack my rucksack and set off for a better land, to follow the stream of tourists part of the way, perhaps as far as the Yosemite Valley, and then perhaps to carry on from there into the Sierra Nevada.

My bag was almost packed and I was all prepared to set off unaccompanied when suddenly I gained a travelling companion in the person of Dr Fr. Ratzel from Karlsruhe. Our unexpected meeting in such a distant country put us both in the very best of spirits, which augured well for the future.

Thus begins Eisen's travelogue 'From California to Nevada', a supplement in the journal *Land and People*, published in Stockholm in 1876–7 by the Society for the Dissemination of Useful Knowledge. The subtitle of the hundred-page narrative is 'Descriptions of Nature', a title that truly fits the bill.

First of all a few words about Ratzel, after whom Eisen had actually named a worm a couple of years earlier. They bumped into one another purely by chance in a restaurant in San Francisco. Ratzel had finished with worms, was now working as an international correspondent for the *Kölnische Zeitung*, and he had just completed a year-long journey around the United States and would soon be on his way to Mexico. But the mountains sounded interesting, and Eisen was burning with enthusiasm for civilized company at last.

Friedrich Ratzel (1844–1904) is a major figure in the

history of science. A number of his books are still read, and even today his name is respected among those involved in cultural geography. The very fact that he had started as a biologist before his interests moved on into other fields gave his theories a special cachet and permanence, though it also brought its own risks.

The titles of his first three books tell us something about the road he took: *Wandertage eines Naturforschers* (*Wanderings of a Naturalist*, 1873), *Die Vorgeschichte des europäischen Menschen* (*The Prehistory of Europeans*, 1874) and *Städte-und Kulturbilder aus Nordamerika* (*Profile of Cities and Cultures in North America*, 1876). They tell the story of a young man who starts by digging earthworms and writing about nature and then develops an interest in the prehistory of the people of Europe. Then, at the age of thirty, he crosses the Atlantic with the intention of writing about culture in the United States, with particular reference to town planning. And that's how he continued for the rest of his life. He was enormously productive.

The book about American towns, which is what he was working on when he and Eisen ran into one another, wasn't translated into English until 1988. It's a gold-mine for anyone who wants to understand Americans and their cities, which frequently function well but are often rather boring. Many years later, when Ratzel was professor in Leipzig, he came up with a concept – just a word, really – which would prove to be an undeserved stain on his reputation: *Lebensraum*. Unlucky. It wasn't his fault that long after he was dead others would abuse a concept that initially referred to the distribution of plants and animals and had nothing to do with geopolitics. If only he had realized where that was going to lead, he'd probably have stuck with earthworms.

Eisen was three years younger than Ratzel, and there is no doubt the dynamic German had a great influence on him. For a start, he immediately began writing lengthy newspaper reports that resemble the reports Ratzel was selling to the *Kölnische Zeitung*. Descriptions of nature. It's always possible to sell that kind of thing. I should know: once upon a time *Västerviks-Tidningen* bought some of my thrilling tales from distant lands, so I also understand that the genre demands some degree of dramatization. As a general rule travellers should be taken with a pinch of salt.

But Ratzen's account is certainly interesting. For a couple of months in the summer of 1874 he and Eisen travelled together across bone-dry plains and up into the inaccessible forests of the Sierra Nevada. They travelled by stagecoach, and then on foot, and finally they hired horses. Everything is described in minute detail. Animals and plants and the natural environment in general, but also thousands of other things: historical anecdotes, portraits of gold-miners, Darwinist speculation. Eisen is also smitten by Ratzel's interest in towns, as his description of the town of Merced reveals:

Towns in America do not develop by royal decree as they do at home, but in a much simpler way. An enterprising man builds a pump and by the pump a hotel. A dozen saloons and a 'dry goods store' quickly spring up alongside and the town is born. Emigrants and unsuccessful gold-miners drift in, drink, play cards, quarrel and talk bar-room politics. The community is flourishing! Everyone is trying to earn as much money as possible in as short a time as possible so that they can return home with money in their pockets and play the gentleman. Fortune is not as kind as they had expected and purses do not fill up as fast as is desirable. One or two of the citizens grow weary of their bachelor lives and if there should happen to be a woman available there is nothing to stop them getting married. Mrs X becomes the queen of the town and contributes to the refinement of society. The Chinese flood in from every direction, live in shanties, do the laundry and the ironing, chatter and shout. In short, nothing is lacking and the new community can proudly consider itself to be a place of significance in the great republic. Merced is that kind of place.

This is just one of the many occasions in his reports when Eisen touches on the unhealthy relationship between Americans and money. The whole narrative is characterized by a high level of biological knowledge; and then there is the friendship between two young men, both of whom will become famous; but it is only when he voices his thoughts on the profit motive that we begin to get a more rounded portrait of the author as the fairly well-heeled son of a wholesale merchant and as a man with good prospects in the academic world: 'In general the half-civilized life of America is not to my taste. No art, no poetry, no memories (and these, after all, are sometimes the best things), not even freedom; just the empty reality of conflict, wrangling and dollars. One has to go to nature in order to feel at home.'

Not even when their journey has taken them all the way to Mono Lake, the remarkable volcanic crater in the desert east of Yosemite, where hardly any biologists had been before, not even then can he resist telling a little moral tale about greed. It's the story of the sporting rifle discarded in the middle of the desert to lighten the load on a horse suffering the great distances between water-holes under the desert sun. It sounds like an urban myth, but that doesn't concern me – it the narrator I'm in search of.

The story tells of a second man riding the same way and finding the rifle. It's a valuable Mauser, not the kind of thing you walk past, so, pleased with his find, he takes it with him, because he is only on day one of his desert journey. But the going gets more and more difficult, and after a couple of days his horse is exhausted, his water and provisions running out, and his only desire is to get out of the desert alive. He throws away the rifle and survives.

Sometime later he returns the same way and hunts in vain for the place he ditched the rifle. It's not there. But a few days later, deeper in the desert, he finds it in the sand. 'It had clearly been found by someone else who had gone on to have the same experience of weight and distance and been forced to throw it away. The story goes on to tell that the same rifle has travelled back and forth across the Nevada Desert for twenty-five years and it's said to be still on the move, though somewhat the worse for wear.'

This story probably exists in all cultures, albeit in different versions. I like it, although I'm well aware that its moral – that wealth is a burden – is most likely to be heard on the lips of people who are already comfortably off. Eisen's own good fortune would soon be a thing of the past.

. . .

It was already swarming with people when he arrived at the Stock Exchange on a sunny June afternoon. A dazzling assembly: statesmen, geniuses, scholars, military men and civil servants of the highest rank. There were uniforms, academic garb, the stars and sashes of numerous orders, all gathered together with one great common interest: the promotion of that humanitarian institution, marine insurance. It calls for great love to hazard one's money on behalf of those in distress as the result of some disaster. And here was love – Falk had never seen so much love in one place at one time!

The Red Room (1879) is one of the funniest books written in Swedish. Strindberg sees through all the game-playing and grand spectacles and focuses on mankind's deepest passions. In the chapter 'The Triton Marine Insurance Company' he transforms the ancient dream of unearned income into black farce. A speculative bubble. There are occasions when time seems to have stood still.

In the real world the name of the company in question was the Marine Insurance Company Neptunus, and when it went bankrupt it emerged that – for reasons known only to his wholesaler father – the whole of Eisen's inheritance had been invested there. No more than a paltry thousand kronor could be salvaged, and Eisen donated 800 kronor of that sum to Stuxberg, who doubtless drank it away pretty quickly in the company of happy friends.

I've seen it happen before. Benevolence is not always the easiest of things to practise.

People who prefer giving to taking, who smooth the path for others, are not necessarily remembered for doing so. More often than not, unfortunately, unselfish people run a greater risk of being forgotten. When Stuxberg wanted to name a centipede after him, Eisen firmly turned him down. He thought their friendship was too close and that his motives for collecting might be misinterpreted. 'There is nothing,' he wrote, 'uglier than self-glorification.' If nothing else, that tells us quite a lot about how highly he valued the little creepy crawlies he found by poking around in tree-stumps or turning over stones. And they cost nothing.

Now that his inheritance was gone with the wind,

he didn't have any money for the journey home anyway. Consequently he was soon afflicted by that typically American disease, the visible symptom of which is the urge to make a fortune in as short a time as possible so as to become economically independent and thereafter in a position to do all the other things in one's own good time. Thus, parallel with his career as an earthworm expert, which can scarcely have been a great earner, Eisen turned his hand to viniculture. In Fresno.

Chapter 11

The Subterranean Garden

We flew to San Francisco on a day in September and immediately rented a Ford Mustang at the airport. A sky-blue V-6 cabriolet. We dropped the hood and, under a baking sun, drove east along dead-straight roads towards the mountains. We had the stereo on high.

. . .

I'm not stupid. I know all about limits and boundaries. In our present age, with the climate crisis accepted as an absolute truth virtually every time people talk about the future, this sort of behaviour is unacceptable, unless you want to be remembered as an agent provocateur, that is. I realize that, but I don't go along with it. So let's talk for a moment about conviction. Faith? We'll get to that later.

The issue of the environment – forests, seas, climate – has become a religion, albeit a secular one based on science rather than on ancient stories of gods. It's a good religion, better than any of the old ones, and no one is happier about that than I am. The spiritual needs and eternal striving of mankind to behave decently to one another have always sought new paths, new narratives. This is one of them. Communism was another and, in principle, it wasn't stupid, though it all eventually went awry.

I don't know what the best approach should be if we want this new path to work. My sole desire, however, is for tolerance – religious freedom, if you like. There must be space for doubt to exist without its being opposed with the viciousness of the Inquisition. It was relatively easy to remain united on the issues during the first decades of environmental discussion because scepticism is the life blood of real science. But now that things have moved on to the evangelical phase, there is a demand for doctrinal purity and for everyone to believe the same thing. That is not a good thing.

As I said, I don't have a particular prescription; all I have are my doubts.

I'm not fully convinced by the climate warnings. To an extent, perhaps, but essentially no. Since I've been following the subject very carefully for many years, I could produce scientific reasons for this, but I'll refrain from doing so. I'd rather just say that I don't believe it: doubt, after all, is at least partly a matter of temperament, not just of knowledge. I doubt if I would have been a very good Communist either, and as for gods, well, I've never believed in them.

There's nothing the matter with religion per se. It's the institutional forms that I find disquieting, the irreconcilable polarization that means that the words I am permitted to voice here between the two of us – between you and me – are said to be unethical and provocative if I voice them in the public space, where doctrine demands that the sceptic should be ridiculed and driven out into the desert among the real idiots, the real enemies of the planet. As if everything were simply black and white.

I find nothing more boring than things that are designed to be provocative. It may sometimes be acceptable in private life, but when it comes to the art and politics of the present day, I simply feel weary when people try to provoke me. I no longer get angry when people who want to change the world or who want to attract my attention for any reason whatever presume that I am so satiated, so anaesthetised by advertising and images, that I need to be given a good shaking, need to be upset and frightened with exaggerations that are not very different from medieval ideas about purgatory. As if that were the only way forward.

I want to be taken seriously.

That is why I approved of Al Gore. His honesty.

I don't actually share his belief that carbon dioxide is the evil spirit of our time, but we are adherents of different strands of the same religion, him as a preacher on the stage and me by temperament more of a doubter in the wings. So back to the Mustang: I could, of course, think tactically, play politics, hold my tongue, not stir things up and simply pretend that we would much rather have cycled through California and we took the Mustang only as a necessity, as a necessary evil. That, however, is not a pew I have any desire to occupy.

. . .

The woman behind the counter at the car-hire place in San Francisco could see that we needed a Ford Mustang. A sky-blue V-6 cabriolet. Her eyeshadow reminded me of the neon-tetra fish I had in my aquarium as a child. She smiled and exhorted us to have a nice day. And then we played Frank Zappa on the stereo in the sun.

Hey there, people, I'm Bobby Brown ...

The travelogue Eisen published in the journal *Land and People* is among the best things he wrote. They had a great time together, he and Ratzel, that's obvious. He had an inquisitive eye and almost everything was new to him. Look, a roadrunner! The bird that can leave a galloping horse standing. The descriptions expand and extend, the stories link together.

Woodpeckers, owls, the fungus *Helotium aeruginosum* in caves. And the trees, of course. The Sierra Nevada glories in wonderful trees. Oak, spruce, pine, every variety. And the peerless mighty sequoia is unforgettable. The journey Eisen and Ratzel make to see them, the world's biggest trees, is effectively a pilgrimage.

Yosemite, however, is the only place he finds impossible to describe.

> *The true magnificence of the Yosemite Valley, with its vertical granite precipices, its cascading waterfalls, its tall dark forests, can hardly be done full justice, since even the most vivid pen can offer no more than a pale approximation of the wonders of this natural magnificence. If I now proceed to describe the main features of the valley and its nature, it is merely to provide the reader with a framework for his own imagination, the boldest flights of which will come closer to the reality than my pale description.*

From time to time my work puts barriers in the way of a normal family life, but whenever it proves possible to combine the two everything is much more fun. Johanna loves fast cars, and we had both come to regret our decision not to visit Yosemite when we had been criss-crossing Arizona and the adjoining states tracing the story of Gunnar Widforss. Widforss may have made his home in the Grand Canyon and that was where he died, but it was in Yosemite that his career as painter of the national parks had begun.

Eisen is also buried in the mountains, but in a different park further south, in a paradise that he managed to save for posterity. So I had another pilgrimage to make. Yosemite was the start, just allowing a couple of days and a couple of nights to get our eyes accustomed to the beauty and the scale, so to speak. We drove straight there, stopping only very briefly in Oakdale, a one-horse town on the plain, to take a rest and to study a collection of barbed wire they have in a museum there. They had a variety of makes from 1864 on, including a piece of electric fencing from 1918. There's always something new to be learned.

I was in unusually good spirits, possibly because my aversion to travelling had receded a little since I'd given up hoverflies. Now that I was collecting only the genus *Callicera*, everything felt meaningful again, probably because it made it possible to survey the field properly. During our weeks in California I kept my eyes open the whole time – in the mountains, around Fresno, Cambria, Monterey and San Francisco – and saw nothing, nothing at all. That's what I'd expected: American *Callicera* species are so rare that they make you wonder whether they exist at all.

We descended into the valley in the evening and checked in at the Ahwahnee Hotel – opened in 1927 and all wilderness-kitsch – mainly to have the chance to view the Widforss watercolours that were hanging in the lobby. Some of them were very good, others not so good. That, too, was expected.

The Yosemite National Park was founded in 1890, one week after the Sequoia Park, but even on Eisen's first visit the valley was already a tourist attraction, though only for the few adventurers who had well-developed leg muscles. Several million visitors a year go there these days and they are right to do so. We hired bicycles and rode around under the late-summer sky and all I can say is that the park is as divinely beautiful as everyone says, or tries to say it is. It's not something you talk about unless you have to. If the beauty of the Grand Canyon is brutal and sometimes simply big, the beauty to be found in Yosemite is as harmonious and bright as the recessional hymn at a school-leaving ceremony.

Fresno, however, is quite the opposite.

This city of half a million down on the plain was the destination of the first part of our journey: not long ago it won – against strong competition – the title of 'Least Liveable City in the US'. That stirred our interest. As part of the process the mayor was asked to state the best thing about Fresno. There must surely be something good?

He thought for a while and then said that the fact the temperature rarely goes above 120° Fahrenheit must be considered an advantage. For those more familiar with Celsius, that's 48.9°. The evening we arrived was a borderline case. Being Europeans, we stuck to our old habits and aimed for the centre. All we needed was a restaurant and a hotel.

I have never experienced anything like it since the curfew in Liberia in 1981. We turned off the highway in the twilight and parked in the middle of town. It was lucky there were two of us, otherwise I'd have felt like the hero of Jens Sigsgaard's classic children's book *Palle is Alone in the World*. No restaurants, no hotels, nothing but closed office-blocks. The only people we saw as we took a longish walk through the heart of the city were some rough sleepers and a number of shabby invalids who emerged here and there in electric wheelchairs. It was like a horror film.

. . .

Eisen endured some hard times in Fresno. He remained in contact the whole time with the California Academy in San Francisco and with earthworm friends the world over, but research funding soon dwindled and he was forced to support himself in other ways. Since his half-brother Francis had bought a farm in Fresno as an investment rather than with any intention of working it, he handed over the running of the place to Gustaf, who was a good twenty years younger than him.

The first time they rode into Fresno, Gustaf counted all the buildings he could see. There was a total of seventeen, and that included the stables and smallish sheds. The town on the plain was so new that the Eisen brothers would come to be numbered among the founders and in time the Eisen Vineyard would become a model enterprise. They started with just a few acres of vines and a small field of alfalfa as fodder for the horses. Five years later, around 1880, the farm covered 300 acres and had its own dam for irrigation, as well as a distillery and extensive fruit orchards. The wine production was approaching 110,000 gallons a year.

Eisen's collecting habits also turned in new directions. He planted a rose garden that eventually held 140 different varieties, and he had a private zoological museum that contained lizards, snakes, frogs, mammals and just about anything else you can think of. In his letters to Stuxberg he writes about growing bananas, olives, cotton, orchids and narcissi. He imported vines from all over the world and claimed he had a hundred different varieties of grape under cultivation.

He was always experimenting, at first with tobacco

in particular, since that seemed a promising crop. He and his brother-in-law planted twenty-five acres but the climate proved too dry and it didn't make them any money. Raisins turned out to be a much better bet.

The idea originated with one of his friends in Fresno, a Catholic priest called John Bleasdale, an Englishman who had been resident and active as a priest in Australia for many years. Bleasdale was also a corresponding member of the Linnean Society and a cultivator of a number of varieties of grape. The Church authorities in Australia, however – an extension of the arm of the Curia perhaps – saw to it that he was moved out and on to California, because it seems that Father Bleasdale's leisure hours were spent drinking hard in the company of the Archbishop of Melbourne, who had alcohol problems. Bleasdale landed up in Fresno and it was as a result of his advice that Eisen began importing from Australia the varieties of grape most suitable for the production of raisins.

The first commercially cultivated raisins in the San Joaquin Valley came from the Eisen Vineyards. What may look like a stroke of luck was almost certainly a result of Eisen's inquisitive mind and scientific background. As with the tobacco, he suffered many setbacks in the early days, but he plugged away and got on top of the problems. After a couple of years he bought land of his own in Fresno and started a nursery, Fancher Creek Nursery, and during the 1880s he became a leading figure within the California horticultural industry. He was the gardening correspondent for a range of newspapers and journals for many years, and he wrote literally hundreds of articles.

He might easily have disappeared from our view at this point as just another emigrant, but his plans were substantially grander than those of an ordinary farmer. He may have been driven by a desire for fame, or maybe it was just restlessness. From early in the 1880s he began to undertake long journeys – Guatemala was the start. With the passing of the years the grower in him became more and more of a theoretician. No one knew more about the cultivation of raisins than he did, but, instead of staying and becoming a wealthy raisin king, he sold his land and gathered all his knowledge into a book, *The Raisin Industry* (1890), which covers all aspects of the raisin from its cultural history and appearance in literature (Shakespeare's *Henry IV*) to recipes for raisin jam. He repeated this tour de force in a later book about the world of the fig, *The Fig* (1901). But we are jumping ahead of ourselves.

Or perhaps not. To some extent Eisen's career as an expert on methods of cultivation runs parallel with zoological research in San Francisco. The subjects he chose were narrow and his writings call for readers with special interests. A simple chronology does not tell the full story in his case, because there are too many dark corners.

Why he left the vineyard in Fresno and went to Guatemala at the start of the 1880s, for instance, is a mystery. He himself gave different versions of what happened. Officially he needed a change of air in order to shake off a touch of malaria, but it seems that a rather more personal reason is to be found in disagreements about the best way of running the farm.

I believe he was depressed.

'At present everything looks pretty black.'

His friendship with Anton Stuxberg was long since over, although he wrote one last letter in February 1881. It's a sad letter. He'd heard from others that Stuxberg had just got engaged and was at last in a financially rosy position. Eisen himself was 'in very bad circumstances', and for the very first time in all these years he was the one asking for money. He didn't even get a reply. My belief is that he simply wanted to get away from it all. He was approaching the age of thirty-four and what had he really accomplished?

Somehow or other he had become acquainted with a deposed Central American president – I've no idea which one – who was living in California and plotting how to regain power back home. Eisen seems to have been drawn into this game, possibly hoping to make some money. Whatever the reason, this man provided him with high-level contacts in Guatemala and that might have been another reason why he went there. It's all rather unclear, but in January 1882 he took the boat south from San Francisco.

. . .

Far and wide, far and wide. For more than a year Eisen criss-crossed Guatemala, mostly on foot, and as usual he collected everything he came across. The expedition is described in detail in three long travelogues published in Stockholm in 1886 and 1887 in *Ymer*, the journal of the Swedish Society for Anthropology and Geography. Ten years later his essay 'A Journey to Baja California and Sonora' was published in the same place.

I mention this with a sense of shame.

These reports, the last things that Eisen wrote in Swedish, leave me cold, and I feel ashamed of myself. It's as if I weren't being sufficiently committed or, even worse, as if I were merely using Eisen as a sort of scaffolding for my own story, which, I'm afraid to say, interests me much more. This insight, or the mere suspicion of this, is unbearable.

But what can I do about it? There was only one single passage where I felt fully involved.

> *At dusk the whole of nature seemed full of life and fire and a host of fireflies buzzed through the air. At one moment they suddenly flared up like sparks of electricity,*

at another they hovered forward, their lights fading. The biggest of these fireflies – 'lucernas' as they were called here – was a large beetle (Elater) several inches long which shone with a light on the outer edge of each wing-case and with another much yellower light underneath. I collected a great many of them in my net, which, when I arrived home, I attached to my hammock with a cord and they really lighted me to bed.

Ford bugs, that's what these beetles are called in English, the reason being that they look like cars at night, complete with two forward-pointing lights and a more yellowish rear light. I've never actually seen them but I did once face the problem of translating the name Ford bugs into Swedish. After a lot of thought I came up with a solution – Raggarbugs, after the Swedish *raggare*, retro greasers who favour classic American cars.

I don't take on translations these days. I do still have a sense of shame, though, about my own butterfly mind among other things.

. . .

Fresno is semi-desert at the start of autumn. Eisen describes the area as a steppe that could bear fruit only with the help of intensive artificial irrigation. The countryside is not much to look at, and Eisen's own landholdings were turned into a golf-course even before the Prohibition of the 1920s made many vineyards unprofitable.

We found a hotel eventually, and the following day we roamed around the fringes of the city in the hope of finding some of the eucalyptus trees said to have been planted by Eisen, but the heat was so intolerable that we avoided getting out of the car. And, since our visit to the city seemed pretty pointless in every way, we decided to go back up into the mountains. I had just settled back in my seat with the map that covers the road to the Sequoia National Park some seventy miles to the southeast when we made a sudden change of plan.

Johanna caught sight of a sign and stamped on the brakes right in the middle of an industrial zone that looked unusually dreary even by local standards. As if the traffic itself wasn't enough of a problem, she had taken to reading the billboards at the roadside and she had been struck by one that proclaimed 'Forestiere Underground Gardens'. A smaller signpost informed us that a guided tour was scheduled for half an hour's time.

The place turned out to have a story to tell, and, although I'm not absolutely sure, I think it says something about Fresno. And maybe something about exile and loneliness, too.

Baldassare Forestiere (1879–1946) was a poor immigrant from Sicily who arrived in New York at the beginning of the century and toiled there as a manual labourer for some years before the dream of California drew him west. He wanted to become a fruit-grower, like his father. The price of land was still cheap and the advertising was seductive, so he invested all his savings in a few acres outside Fresno in the San Joaquin Valley. He travelled full of hope – but woe the man who trusts to fortune!

The land turned out to be worthless. The surface soil seemed fertile enough, but just six inches down there was a crust as hard as concrete – known as 'hardpan', it was a geological phenomenon not unusual in the area. Fruit-farming was out of the question, and young Baldassare had no money left, just the piece of land, so he built a wooden shack and survived as best he could by working as a day labourer on other people's ground.

The heat was killing. What was he to do?

Maybe it was homesickness that made him recall the sweet coolness of the wine-cellars of his childhood in Sicily. He started to dig. Once he had worked through the hard crust, everything became surprisingly easy, and within a short time he had an excellent cellar in which to store his food and his cheap wine. It was pleasant and cool below ground, so he carried on digging and excavated a small kitchen and a dining room. Later he added a bedroom, and then a well and a fishpond.

Next he discovered that it was possible to grow things down there, a couple of yards beneath the surface. The soil was fertile, the moisture stable, it was cool in summer and mild in winter. All he needed to do was to construct well-like shafts for the fruit trees to grow up through. Spurred on by his discovery, he carried on obsessively burrowing away year after year. He kept at it for forty years, his whole life, and in the end he had created a fantastic labyrinth of winding tunnels and walled vaults. At its deepest it went down no less than twenty-four feet and the total area covered 1,200 square yards.

Johanna and I were the only people on the tour. A grandchild of Forestiere's brother guided us round the labyrinth, one of the most beautiful homes I've ever seen. Without all the trees, the lemons and all the rest, the catacombs would undoubtedly lose much of their beauty, but what we now saw was a paradise garden, with vines and flowers growing everywhere and trees standing in the middle of mostly circular rooms with just the treetops above ground. Light filtered down through the leaves as it does under a maple tree in May. Big rooms, dark chambers, fountains, kitchen gardens, here a pergola, there a chapel, a swimming pool – a whole life.

'Did he ever marry?'

The guide seemed surprised by the question.

He fidgeted a bit and then said no, not as far as anyone knew. There had been a woman but she had left him. They had been engaged in their younger days. There was a moment's silence. Johanna didn't ask him any more than that but he continued, a touch hesitantly, and told us that Baldassare had always hoped she would come back. It was for her he had created all this beauty, but she never returned.

Chapter 12

The Oaks around Gränsö Castle

Eisen doesn't seem to have been much of a man for women, or if he was he hid it well. It's none of my business, of course, but I did wonder sometimes. Love can explain almost everything in someone's life.

My imagination took off, but nothing came of it. And, anyway, there were many other things that remained hidden and inaccessible. I had long since given up the idea of presenting Eisen's life as a straight line with a beginning and an end, when one day I happened to hear of a financier in Los Angeles, by all accounts a very successful one, whose publicity material claimed he was a direct descendant of Gustaf Eisen – the celebrated scientist and collector, as it puts it in the blurb.

The American finance market at the time was suffering its worst crisis since the Depression of the 1930s, so I had few illusions about being able to contact the man in question. I tried anyway and it seems the firm wasn't too busy because within a few minutes a whole battalion of public relations men were on to my case, though none was able to provide a proper answer to my question. The financier himself, the owner of the company, was unavailable, but they promised to pass on my enquiry.

The very next day he got in touch with me. There is nothing that interests Americans more than family business and the story turned out to be essentially true. It depends on what you mean by a direct line of descent: he proved to be a fifth generation descendant of one of Eisen's half-brothers, an architect in San Francisco. But it was something to be going on with.

We corresponded for a while and exchanged information, and once I became better acquainted with this friendly man I felt bold enough to ask him whether he happened to know anything of the women in Eisen's life. There was some delay before I received an answer – he had presented himself to me as the family historian, so I assumed he needed to do some research before responding. It's quite a sizeable clan and there are many people for him to ask. Eventually the answer came and, perhaps typically, it came as an afterthought: '… the romantic angle by the way. There was a woman he was very close to. Alice Eastwood – she was the one who brought his ashes from New York to his grave in Sequoia.'

Alice Eastwood! I'd heard of her. She visited Sweden once for the major botanical congress held in Stockholm City Hall in 1950. She was honorary chairwoman, ninety-one years old, and a legendary figure. What a find! I immediately searched through the records for any certain indication of true love, and, sure enough, there it was. Eisen had named an earthworm after her: *Mesenchytraeus eastwoodi* (Eisen 1904).

. . .

Eisen's collections were almost completely destroyed, a life's work going up in smoke when the California Academy burned in the chaos that followed the earthquake of April 1906. Very little could be saved, no more than a fraction of the whole, and that little was eventually described in an article, 'On Some Earthworms of Eisen's Collection', published in 1962. A couple of dozen worms preserved in spirit, several of them of hitherto unknown species.

It must have been as if a great bundle of manuscripts were hurled down from the top floor into a burning library. Not even a thousandth part survived.

But, as in all earthquakes, then as now, in the midst of the inferno there comes a tale of miraculous rescue, an epic of heroism, classical and beautiful, about someone prepared to hazard their own life to save what could be saved. That someone could have chosen to save their own home from the flames, but instead that someone chose to climb up the banisters in the museum for six floors, pursued all the way by the fire, in order to rescue the botanical type collections. Alice!

. . .

Alice Eastwood (1859–1953) came from a humble background in Toronto in Canada. Her mother died when she was six years old and her father sent her to live with relations. She was lucky enough to end up with an uncle with botanical interests and he immediately began to teach her the Latin plant-names. Botany was soon her world, and while still a teenager she demonstrated unmatched perseverance out in the field. She would stay away for days, all alone.

She was effectively self-taught, which did not prevent her being handpicked for a post at the California Academy in San Francisco at the start of the 1890s. They weren't too worried about formalities out there, and, unlike other institutions of learning in America, they tried to attract woman researchers from the start. She worked her way up quickly, and at the age of thirty-five became head of the botanical section, a post she held until she retired at the age of ninety.

Her action after the earthquake, when she saved some 1,200 herbarium sheets containing irreplaceable type specimens, made her world-famous, but long before that she had won herself a strong scholarly position, partly by inspiring hordes of amateur botanists – mainly women – and partly by political campaigning for parks and other green areas in the cities. She was also an early advocate of environmental protection and, ironically, from 1903 she was one of the highest ranked botanists in the annual list entitled *Men of Science*.

Eisen was also a curator at the academy. He was in charge of invertebrates, so they would certainly have had a good deal to do with one another, though not many people think they were a pair in the modern sense. But how should we know?

As a woman she was occasionally asked why she had never married, and on one occasion she responded rather caustically that something as impractical as marriage would probably have got in the way of her first love, which was botany. It seems unlikely that anyone ever asked Eisen that question. Anyway, I'm certain each of them would have recognized themselves in the other, for both had experienced a hard childhood and had created their own worlds as collectors at an early age.

They were friends, no doubt close friends, but they were both solitaries. Eastwood was fired by a passion for botany and as for Eisen, I'm not sure I've really managed to work out what he was fired by. I don't think it was earthworms. Collecting per se, perhaps, or systematics. Starting afresh and learning something new all the time is, in its own way, a sort of passion, I suppose. He was, in the very best sense, like a child: fickle and inquisitive. His love for the ancient trees was probably the only thing that burned with a steady flame throughout his whole life.

I have no difficulty in understanding that. Explaining it, however, is more difficult and it may be worth our while, perhaps, to approach it via a circuitous route, via world literature.

. . .

Everyone knows that Astrid Lindgren celebrated one of her greatest literary triumphs when she invented what she called the Lemonade Tree in her book *Pippi Goes on Board* in 1946. Some of the later stories from her pen didn't appeal to me anything like as much, but that didn't stop me loving her and it was mainly because of this tree.

A perfect hiding place, with enough room for

everyone, Pippi, Tommy and Annika. It's said to be modelled on an enormous hollow elm tree, which has since become a major pillar of the Småland tourist industry even though it's an oak tree in the story, a giant oak in which they can sit and idle away their time or spy out through a crack in the bark. And, as if that wasn't enough, at suitable intervals lemonade and chocolate cakes would materialize from nowhere in this magic tree.

I was utterly fascinated by the story. The only thing I found slightly disconcerting was that the children didn't seem in the least interested in the beetles that must have lived inside the tree. On the other hand, that's the great thing about good books – you have to do some thinking for yourself, which was not difficult in this case given that I'd grown up with trees like that.

There were numerous huge oaks, mostly on the other side of the Gränsö Canal, and some of them were rotten and hollow enough to hide in for a while, or indeed to occupy more permanently as secret hiding places in the warm summer nights when time and a good torch were the only things you needed to find the click beetles that live in hollow trees. All you had to do was wait and sooner or later something exciting was guaranteed to turn up.

Astrid Lindgren's readers will no doubt remember that a little procession of beetles does eventually come strolling along in the third book, *Pippi in the South Seas* (1948). This was when Pippi invented the word *squeazle*, which no one understood – including her – until after much laborious searching she and her friends discovered that a *squeazle* is the kind of beetle they see crawling across the yard at home.

It's wonderful. They have searched and searched far and wide, but the answer was right in front of their noses at home the whole time.

> *'Oh, look at the beetle!' Pippi yelled.*
>
> *All three of them crouched down to look at it. It was so small. Its wings were green and looked like metal.*
>
> *'What a pretty little thing,' Annika said. 'I wonder what kind it is.'*
>
> *'It's not a cockchafer,' Tommy said.*
>
> *'Nor a dung-beetle,' Annika said. 'Or a stag-beetle. I really wish I knew what sort it was.'*

A blissful smile spread across Pippi's face.

'I know what it is,' she said. 'It's a squeazle.'

If we take things a step further and pretend to be literary critics, the whole story of the Lemonade Tree could be viewed as one of the most ingenious metaphors in Swedish literature, with a structure like a three-stage rocket. The same tree actually turns up in Lindgren's first book, *Pippi Longstocking* (1945), where it figures as the famous hollow 'thing-finder' tree-stump in which Tommy is amazed to find a notebook with a silver pencil and Annika a red coral necklace.

'You can practically always finds things in old tree-stumps,' said Pippi.

It seems that these books, which were read aloud to me at an early age, made such an impact that even now I lose my critical faculties when faced with anything to do with old hollow trees. Ernest Callenbach's novel *Ecotopia*, a bestseller in the 1970s, holds some attraction for me for the same reason, even though, viewed in a sober and critical light, it is pretentious twaddle.

It tells the story of an ideal future society on the American West Coast. Northern California, Oregon and Washington State have broken away from the United States and distanced themselves from the world around them in order to create an ecological paradise, though its isolation is reminiscent of North Korea. Will Weston, a journalist, succeeds in getting into the country and soon comes to the Sierra Nevada, where he falls in love both with the sequoia trees and with the free-spirited ecotopian Marissa:

> *We came to where the forest was denser. She suddenly disappeared into a hole in the back of a gigantic tree. I ran in after her and discovered I was in some sort of shrine. She lay there on a bed of pine-needles and her breathing was deep and breathless.*

I don't think we need to go into further detail at this point! The environmental movement of the 1970s had its particular attractions; we can call them social: it wasn't birds and insects that attracted most of the supporters and often it wasn't environmental issues at all, but rather more basic instincts. You don't have to be a cynic to recognize this. The same lusts and longings for togetherness have always underpinned political and even religious movements. Nothing strange about that and usually no great bother. Rather the opposite if anything.

The love for great trees is age-old, but not in the United States. In Europe there exists a pre-Christian tree-cult that the Church never fully succeeded in eradicating, but in Eisen's day the settlers in the New World were still in the grip of a lust for profit that accepted the ruthless felling of ancient forests. The redwoods (*Sequoia sempervirens*) along the Pacific coast were the first to go and then the pioneers moved up into the mountains, where the even mightier giants grow – *Sequoiadendron giganteum*.

The redwood forest on the coast was extensive and had been known since the middle of the eighteenth century, but the true giants, which were not discovered until 1852, only grew in scattered small groups up in the Sierra Nevada. They had been rare right from the start. And, as I said, they were big – quite how big is difficult to describe. When he was director of the Museum of Natural History in Stockholm, Eric Hultén shipped home a cross-section of a storm-felled sequoia's trunk; it is now in the foyer there and it has 2,400 annual rings and a circumference of forty feet. It's impossible to walk past that great slice unmoved, and the mere fact that it was sawn by hand stops you in your tracks. There is another story, however, that is frequently told to describe the size of these trees; a very American and profoundly symbolic story of the greatest tree destroyed by man.

When it was chopped down its circumference was ninety-five feet. This happened in 1853, shortly after the gold-rush, and on the stump they built a dance floor with space for sixteen couples at a time. And for a decent-sized band.

We went to see it, of course.

The abundance of fresh bear tracks on the path that morning had convinced us that we would be better viewing Mount Eisen from a distance. Actually walking there seemed unnecessary and, since the mountain is 12,185 feet high, it can be seen in the distance from all around. We still hadn't located Eisen's grave, which is a simple cross, though heaven knows where, so instead we spent a couple of days lazing on the terrace of the hotel and enjoying the fresh air. After our experience of the desert in Fresno we had taken up residence at high altitude in the national park.

There was frost on the ground the morning we walked down to General Sherman. Yes, that's right, the biggest tree in the world is called General Sherman.

And what can I say about it?

Not much more perhaps than that on the short stretch from the parking area a couple of Italians walking in front of us chattered away and gesticulated as Italians do, but suddenly lowered their voices and began to whisper when the colossal tree came in sight. It was just like the way people behave when entering the portals of St Peter's Basilica in Rome. The tree is enormous and has to be experienced on the spot. The volume of timber in this single tree is estimated to be 52,500 cubic feet – like a tower block. If General

Sherman were to be felled, an even bigger dance floor could be constructed, but these days it would be seen as a crime as unforgiveable as the fanatical Taliban demolition of ancient statues of Buddha in the mountains of Afghanistan.

Even John Steinbeck found it hard to describe the giant trees, but he did eventually solve the problem. The piece is to be found in his book *Travels with Charley* (1962), an account of the author's journey around the United States in a camper-van with only his dog Charley for company. Steinbeck was old, famous and rich, and, although he'd been living on the East Coast for many years, he was feeling the call of the land of his childhood – California. That's what had brought them to the Sierra Nevada and now the dog needed to pee.

> *It seemed to me that a Long Island poodle who had made his devoirs to* Sequoia sempervirens *and* Sequoia gigantea *might be set apart from other dogs – might even be like that Sir Galahad who saw the Grail. The concept is staggering. After this experience he might be*

translated mystically to another plane of existence, to another dimension, just as the redwoods seem to be out of time and out of our ordinary thinking. The experience might even drive him mad.

Much has changed over the past hundred years. The American love of great trees now seems to be up there with that of the Germans, a quasi-religious worship with John Muir and Thoreau as its prophets, but that also presupposes practical scientists like Gustaf Eisen. Without him the trees would not even be there: they may have survived in Yosemite perhaps, but not down here in the forests that became the Sequoia National Park.

The year was 1890, in which year General Sherman was still known as Karl Marx.

These giant trees are so utterly imposing in their grandeur that the tradition of naming them after famous people soon developed. It was Eisen himself who discovered one of the biggest and christened it General Grant, a name that still stands. But the biggest of all was called Karl Marx, the reason being that a group of utopian-inclined followers of the Danish socialist Laurence Gronlund settled in the area. A sect, then, on the usual American model, and they planned to buy the land from the state, fell the trees and exploit the timber. They needed a source of income, after all. The sect consisted of a couple of hundred souls, called itself Kaweah Colony and was serious about the trees.

Eisen happened to be in the area at the time, acting as adviser to a land-owner with farming ambitions. He had been roaming the Sierra Nevada for fifteen years, either collecting specimens or just relaxing from his work on the farms, and as soon as word reached him of the threat to the stands of sequoia he did not hesitate. No one knows any longer exactly how the creation of the national park occurred, because so many documents disappeared in the earthquake, but active research is being carried out even as I write and everything points to Eisen having acted alone, played for high stakes and won.

There is good evidence for his having lectured in the California Academy on the threat to the trees and for the academy having given him the task of coming up with a plan for a nature reserve and producing a map that could be presented to the bureaucrats in Washington. It's also known that there were powerful economic interests at work behind the scenes. The Southern Pacific Railroad and other companies were after the same forests as the Kaweah Colony, and they had virtually unlimited influence. And the academics in San Francisco were just as much in the hands of the big companies as everyone else.

The map produced by Eisen proved unacceptable, and the area he proposed for the park was cut by half before the document was sent east. On 25 September 1890, therefore, a relatively small Sequoia National Park was inaugurated. What historians have never been able to understand, however, is why President Benjamin Harrison signed off a further decision just a week later – a decision that trebled the area of the park and included within it the land the developers were laying claim to.

I was made aware of this confusing chain of events by a historian at the museum in Sequoia itself, and we remained in contact, each researching the topic further on his home ground. He came to Sweden later, after it emerged that copies of documents missing in San Francisco had turned up among Eisen's papers in Uppsala. Not until the spring of 2009 did documents surface revealing that Eisen was so disappointed and enraged by the slashing of his proposals by multi-million-dollar interests that he drew a new map proposing three times the area and, using his contacts in Washington, sent this in secretly.

Sequoia was the second national park in the United States; only Yellowstone is older. It is almost as big as the Sarek National Park in northern Sweden, and it is only now, 120 years later, that it has proved possible to shed light on its genesis. Eisen himself never made any great fuss about his role in the drama: what was important to him, I think, was not personal glory but the issue at stake and his love of the place. After his death others inflated their own rather more modest contributions at the expense of his, and I, for one, find that maddening.

. . .

Even my giant trees on Gränsö are protected nowadays. I don't really know how it came about, but I suppose it's the spirit of the age. A lot has happened over the last thirty-five years. During the 1970s, when hardly anyone lived out there and the castle stood empty and neglected, great centuries-old trees were felled without a second thought, but that is no longer acceptable. Widespread development has taken place, housing has sprung up, and the castle has been renovated. The overgrown jungle I used to have more or less to myself is once again a disciplined piece of manor-house parkland, and the interesting thing is that the protection of the trees has increased simply because people live there. That those in power might have read Astrid Lindgren as children may also have something to do with it.

The big old oaks around Gränsö Castle that aren't hollow are as solid as concrete pillars. Kit Colfach (1923–2002), doctor, film-maker, war-hero and adventurer, lost his life when he crashed into one of them, and there was little more than a scratch on the bark of the tree, though very little of the car was left.

Every time I pass that tree my thoughts go to him and I hope that someone will tell his story one day.

I can't. He was my father's best friend and he was too close to us. He was as unpredictable as the weather.

He was born in Denmark and started life with a pretty colourless name. He joined the resistance movement during the German occupation, performed some heroic deeds, was imprisoned but managed to flee to Sweden. I assume that Kit Colfach was actually his *nom de guerre* and it really suited him. I have never known anyone else with that level of logorrhoea: the moment he entered the room, which was a pretty frequent occurrence, everything else came to a stop. He was stylish and self-assured, insufferably charming and given to such extremes of boastfulness that it was impossible to know fact from fiction.

His repertoire, however, was quite predictable. If there were strangers in the room, he would fire off a series of stories from the war or from other adventures, frequently the *Kon-Tiki* Expedition, which he hadn't actually taken part in, but so what. If my memory serves me right, he had declined when Thor Heyerdahl, the expedition leader, asked him to join them. Or he would come out with hair-raising insider tales from

Hollywood – he was a film-maker, after all. In his later years he specialized in pointless stories of the luxury yachts on which he spent his spare time acting as private doctor to American millionaires. If we were really unlucky, he would show a film with endless shots of the Antarctic, which is perhaps the cause of my notable lack of interest in penguins.

But he was a divinely talented doctor, a magician who could cure us all. If he had come up with an illness called *squeazle*, we would have believed him without the least hesitation. The Queen of England awarded him the highest medals for bravery that can be awarded to a foreigner, and I saw them on his coffin at his funeral in St Gertrud's Church in Västervik. He is said to have been a saboteur behind the German lines, but that is something he never talked about.

Chapter 13

In the Name of the Immortal Collector

Ordinary people no longer remember Gustaf Eisen. When prompted by leading questions, a few biologists were able to recall his name, but no more than that. Not even professors on the top rung of the earthworm ladder could come up with anything worth reporting, and, as for the Strindberg experts, Eisen played no more than a non-speaking walk-on role. One or two book collectors did a bit better. But, even taking the whole lot of them together, nothing more than a few fragments emerged. He has effectively disappeared. Why did that happen?

There are many ways of sinking into oblivion. In the first place, Eisen lived to such a great age that none of his friends and acquaintances could say anything when he did eventually pass away because all of them had already been dead for years. He was the last. Exile and childlessness didn't make things any better. Added to which is the fact that he died in 1940, when a world war was raging and obituarists had other things to do than praise men like him. He was a fossil. He used to talk about the Crimean War, and one of his earliest memories went back to 1854, when he was seven years old. He was living with an aunt in the vicarage at Harg in Uppland that summer, and for the rest of his life he remembered hearing the rumble of cannon rolling in from the sea as British and French battleships bombarded Bomarsund, the Russian fortress in the Åland Islands.

Age is certainly significant but it's only part of the answer. If he had died in the earthquake and his collections had survived, there can be little doubt he would have been well known today, something of a hero even. Or if he had been murdered by the people who wanted to clear-fell the Sequoia Forest, for he did receive death threats.

He was, of course, handicapped by his habit of starting afresh, of putting his energies into new enterprises. If he had stuck to his metier as a zoologist he would no doubt have had an excellent career, if not at Harvard then back in Sweden, where he would probably have become a member of the Royal Academy of Science. But he didn't do that.

There is something else I've discovered, too, a mere bagatelle perhaps, but it may be relevant in the context: Eisen had a tendency to vary the spelling of his name. An orientalist in Uppsala I contacted, a very learned man, knew of some of his works, but he was amazed to discover that these three people, as he put it, were one and the same man: Gustaf Eisen, zoologist in Uppsala; Gustav Eisen, Californian horticultural expert; and the art historian Gustavus A. Eisen in New York.

A good many of the Americans I've corresponded with have been led astray by this confusion of names.

Not all libraries have managed to get it straight, with the result that discovering the whereabouts of his books is by no means an easy task.

Let's set that unreliable Christian name aside for a moment and look at the uses his surname has been put to. He has had a number of species named after him – worms, algae and other things, too. That's certainly one way of achieving immortality. Since I learned to use the Latin names while I was still a child, I'm quite unfazed by them and actually full of wonder at their meaning and their linguistic melody. Just for fun I've made a list of those to do with Eisen – there's nothing like a list!

Strangely enough there is still no central register of all the species in the world, which is why this list is probably incomplete: it's more of a collection scraped together during many years spent meandering through the scientific literature. I have found a species here and a species there and eventually got round to gathering them, initially just for my own pleasure but latterly with the intention of erecting a little cairn in memory of Eisen. I actually felt rather sorry for him, and it occurred to me that there's nothing that can cheer a chap up as much as a list of the colleagues decent enough to honour his diligence as a collector and systematist in this way.

I have found almost fifty species, five genera and one subfamily, so the following list is no more than a selection. Everything in moderation.

First of all some directions. The names given in brackets are the names of the authors, that is the friends or colleagues who described the species and thus had the privilege of deciding on a name. So, for instance, *Achaeta eiseni* (Vejdovsky 1877) means that in that year Frantisek Vejdovsky (1849–1939) in Prague – he later became very well known – named an earthworm in honour of Eisen.

Achaeta eiseni (Vejdovsky 1877) earthworm
Anopheles eiseni (Coquillet 1902) mosquito
Anthidiellum eiseni (Cockerell 1913) bee
Azteca forelii eiseni (Pergande 1896) ant
Brachystola eiseni (Bruner 1906) grasshopper
Centris eisenii (Fox 1893) bee

Clarkia eiseniana (Kellogg 1877) tracheophyte

Diaptomus eiseni (Liljeborg 1889) copepod

Diplocardia eiseni (Michaelsen 1894) earthworm

Eisenia (Areschoug 1876) genus of brown algae

Eisenia (Malm 1877) genus of earthworm

Eisenia arborea (Areschoug 1876) brown algae

Eisenia eiseni (Levinsen 1884) earthworm

Eiseniella (Michaelsen 1900) genus of earthworm

Eiseniona (Omodeo 1956) genus of earthworm

Eisenoides (Gates 1969) genus of earthworm

Enallagma eiseni (Calvert 1895) damsel fly

Erioptera eiseni (Alexander 1913) crane fly

Eukerria eiseniana (Rosa 1895) earthworm

Fridericia eiseni (Dózsa-Farkas 2005) earthworm

Hermetia eiseni (Townsend 1895) soldier fly

Linyphia eiseni (Banks 1898) spider

Mesostenus eisenii (Ashmead 1894) parasitic wasp

Pardosa eiseni (Thorell 1875) spider

Phacelia eisenii (Brandegee 1891) tracheophyte

Ranunculus eisenii (Kellogg 1877) tracheophyte

Scotoleon eiseni (Banks 1908) antlion

Sminthurus eiseni (Schott 1891) springtail

Tantilla eiseni (Stejneger 1896) snake
Xenotoca eiseni (Rutter 1896) redtail splitfin (fish)
Zophina eiseni (Townsend 1895) horsefly

As we can see from the above, Eisen's friend Tamerlan Thorell in Uppsala came first – as early as 1875 – with a spider, and the list is still getting longer, so in 2005 Eisen was still considered worthy of yet another worm. It's clear he hasn't been totally forgotten.

It is worth noting that both *Eisenia* genera are very early. We've already referred to Areschoug's brown algae, but it's also interesting that the earthworm genus of the same name was established just a year later by August Wilhelm Malm (1821–82), a man who earned an undying reputation by stuffing a stranded whale in 1865: it can still be seen at the Natural History Museum in Gothenburg. The monster weighed about forty-five tonnes, which was impressive enough in the first place, but things got even better when Malm, who had a genius for publicity, turned the whale's insides into a small café where visitors could relax for a while over a cup of coffee and a glass of punsch.

A rumour was soon doing the rounds, however, that the whale's belly had become a popular resort for those with other things than coffee in mind and when a pair of lovers was caught *in flagrante* it was decided the only thing to do was to close the café. That story adds fuel to my long-standing suspicion that natural history studies are frequently no more than a cloak for other, more basic, activities. Or maybe this was just a jolly jape, the idea being to have it off inside the whale in more or less the same way as some daring souls try to do it in aeroplanes in order to qualify as full members of the so-called mile-high club. Simple pleasures.

Since this was the only stuffed blue whale in the world, Malm stood to make a lot of money exhibiting it to the public, but everything points to a degree of excessive ambition having got the better of him. After notable success in Stockholm he managed to transport his magnificent exhibit down to Berlin on a specially built wagon, but once there his company went bankrupt. The earthworm research on which he focused later must have come as a relief. He, too, was in correspondence with Charles Darwin.

But A. W. Malm had made his greatest contribution as a naturalist in the 1850s, long before the earthworms and the stuffed whale, when, by dint of dogged perseverance, he made himself Sweden's greatest expert on hoverflies. His dissertation 'Notes on Syrphici in Scandinavia and Finland' is still worth reading, and in it the reader can share his experience at seven o'clock in the morning of 15 July 1857 when he succeeded in catching a specimen of *Callicera aurata* in a patch of flowering meadowsweet near Torebo Gård on the island of Orust in the Bohuslän Archipelago.

. . .

My early experiences of mild summer nights were profoundly serious. I have never been more of a scientist than I was when I was twelve years old. When I found my first glow-worm in the grass one night in July, it was not merely a sensation, a miracle, but it also led to my going to books to find out what made this yellow-green glow possible. Collection and information were the only rewards I needed and being alone was a purely practical choice. Nothing special about that. Whatever it was that adults got up to in those same nights was of no interest to me at all.

That was then, but before I knew it everything changed. All of a sudden my greatest desire wasn't to catch big ground beetles of the genus *Carabus* and keep them alive in a terrarium. No, it was something quite different. First of all – as a kind of preparatory stage – it was to measure myself against other boys, and then – cautiously – to approach girls. For a while I even played football to make myself stand out. Nothing much came of that.

It was only after a couple of pretty wasted years that I gradually learned that old passions and new could be combined. To some extent, anyway. Insects were clearly a solitary occupation and doomed to remain so. But then I started with birds as well. As a past producer of nest-boxes, of course, I had a certain amount of elementary knowledge as starting capital. But I wasn't attracted by the run-of-the-mill birding expeditions, the ones where you get up at five o'clock and then stand and freeze along with half a dozen other boys hoping to see a little grebe in Lilla Strömmen. Might as well play football. The fun really started with owl expeditions: you shivered like a wet dog then, too, since owls hooted best in March, but the great thing about owls was that girls used to come along as well.

I don't know why, but they did. It didn't take much imagination to work out what nocturnal excursions in summer nights might lead on to. I climbed the career ladder in the Västervik Young Naturalists' Club at record speed, and once I was chairman I had virtually unlimited influence on the programme. Night-time songsters, I said, we should listen to birds that sing at night. I won't deny the element of calculation that lay behind this decision.

The grasshopper warbler and the marsh warbler were both gaining ground during those years, and their songs were new and exciting. All of us had heard nightingales, of course, but we weren't hard to please and even if the information sheet for the meet didn't offer the hope of anything more than the wheet-wheet call of a spotted crake in some bog within cycling distance, a surprising number of members turned up for the excursions all the same. Ten, perhaps, and all hopeful. How we cycled! Nightjar, quail and snipe. A corncrake out on the meadows beyond Segersgärde on a May night with chocolate in our vacuum flasks came close to being an erotic experience, even if it was only in the sense that the hills between Kuggviken and Habors Klint set our pulses racing and raised our temperatures. We sat together in the darkness and listened in silence for a long time.

. . .

All these nights are locked away in my memory. A distant birdsong, or the scent of the lesser butterfly orchid along the side of the ditch, or almost anything, can set them running like romantic films, and even when I'm groaning with age I'm sure my last memory will be that night in Grönhögen. I doubt I'll succeed in really putting over the story, but I'll try.

We had gone all the way to Öland, since we were now of an age when some of us had driving licences. Just four boys: the girls' interest in birdwatching shouldn't be exaggerated, and our longer expeditions, which involved sleeping in tents and living in all-round primitive conditions, didn't have the same attraction as cycle rides in the warm nights of summer. That's understandable. I didn't think it was that great either, but by some irresistible natural law our boyish curiosity had turned into a competition about seeing the most species, which meant that Öland was a good place, particularly in late spring and early summer.

We had been up since dawn and now dusk was well advanced – it was still before midnight, but late. We'd slept for a while in the middle of the day among the trees in Ottenbylund, but now we were weary enough to be walking in silence, in a half-trance, thinking private thoughts or no thoughts at all. We were coming up from the limestone quarry at Albrunna, where a great reed warbler was singing that year and we'd halted immediately north of Grönhögen to round off the evening with a river warbler that was supposed to be somewhere in that vicinity, or so rumour had it.

Thin streaks of mist hung over the fields, but apart from that there was little to be seen. It was dark. You never see the river warbler anyway; you just hear it at night as we did on this occasion. We were on our way back to the car when we heard voices in the mist. We stopped on the road and peered into the darkness with our binoculars. It was almost impossible to see anything, but after a short while we could pick out some people moving slowly out there, wading, so to speak, through a field of rye that came up to their waists.

Two heifers had strayed out into the field and now the farmer was trying to drive them back with the help of his two daughters. Encouraging calls weren't helping and the two beasts kept evading the girls, who, in their bright cotton frocks, seemed to have stepped straight out of an oil painting from the glory days of National Romanticism.

What do you do in such circumstances?

We helped them drive the heifers out of the field and then we stood there by the fence, the girls and us, shuffling our feet in an embarrassed sort of way while the farmer took his heifers home.

Suddenly one of my friends said: 'There's a dance at the mill tonight.'

Birdwatchers have sharp eyes and notice apparently insignificant details. When we had passed Grönhögen on our way north towards Albrunna and the limestone quarry earlier in the day, my friend had read on a roadside poster that the local restaurant, a converted old mill, had hired a dance band for that night. All he said was: 'There's a dance at the mill tonight.'

And since this happened one summer's night purely by chance and without any trace of calculation, all of us were surprised and happy, as if it were the only truly logical conclusion to the evening.

So that's where we went. It was quite close by and we hung up our binoculars and ordered a beer. I danced with one of the girls, although I never thought I would dare to, nor knew how to, especially since I was wearing rubber boots. Not much was said and there was no question of making any approaches beyond dancing. That was all.

Afterwards we used to remind one another about that night, but we never discussed it and it never occurred to me to talk about it to anyone who hadn't been there. I have always thought of it as being an impossible story to tell.

. . .

Even now, three decades later, the bird hides in Sweden are mainly populated by boys, both young and old, though the number of women ornithologists has grown exponentially over those years. The social circle that used to be like the Catholic priesthood in its male exclusivity has become much more pleasant to be part of. I wish I could say the same for fly-collectors: the only thing wrong with the small circle that exists is that they are all men. In that sense we are still in the nineteenth century.

There seems to be something about the combination of insects and collecting that frightens women off. I don't think the insects themselves are responsible, nor the collecting for that matter, with all its fussy pedantry. It seems rather to be the combination of the two. Or perhaps it's the simple fact that women have a good nose for sniffing out the musty scent of past centuries that clings to these solitary men who are more interested in punsch than poetry and who, while no longer in a majority, do still exist.

There's a story here, comic and tragic at one and the same time, which reflects this version of manhood. Or might reflect it, although it doesn't have to be told in that way. It could equally well function as a counterweight to our mention of the way Latin names for lower animals can immortalize the memory of human beings. And it can also work the other way around – someone's name can make an insect so famous and desirable that the species is collected and collected almost to the point of extinction. Let's spend a little time with that rare beetle *Anophthalmus hitleri*.

A name is forever. As long as everything has been done properly from the start, it cannot be changed: international regulations are very strict on that point.

Anything at all can be named after anyone at all. Gods and their equivalent within established religions are effectively the only names that are not acceptable. That's understandable: a pubic louse named after the wrong prophet would be a nightmare. Politicians, however, are fine, which is why one fine day in 1933 a German entomologist in Slovenia got the bright idea of naming a hitherto unknown beetle after Adolf Hitler. It was homage. The beetle was blind, a pale brown ground beetle, hardly bigger than an ant, that turned out to live in perpetual darkness deep in the labyrinth of limestone caves in the Slovenian mountains. But it was homage for all that, and the Führer is supposed to have been very flattered.

That could have been the end of the story. Just a curiosity. But that leads us on to the problem of lonely men who collect strange items.

Today *Anophthalmus hitleri* is threatened with extinction for the sole and simple reason that unscrupulous hunters who wriggle their way through these pitch-black cave systems armed with head torches and tweezers can sell specimens, dead and impaled on a pin, for a thousand euros each to the kind of fetishist who collects old SS bayonets and anything else from an age when the whole of our continent lay in darkness. It's reached the point where the Slovenian authorities have started stationing armed guards at the mouths of the caves. Really.

Chapter 14

A Mysterious Strindberg Painting

Eisen's talent and tenacity as a collector explain his successes as a scientist. In that respect he is reminiscent of René Malaise in the jungles of Burma. He worked his way tirelessly through the terrain, collecting everything he saw wherever he happened to be. Birds and other higher species of animal were the only things he left untouched – for ethical reasons. He considered them to be too intelligent.

There is one letter that tells us everything we need to know. It was written in 1894 by a dragonfly expert and relates to a collection from Mexico that Eisen had sent for identification. The order of insects in question is a relatively small one, with not many species, in spite of which the batch contained 1,500 specimens, spread across fifty-three species, of which eight were unknown. The writer of the letter would later name a damselfly after Eisen, who doggedly continued to send parcel after parcel of animals and plants to colleagues all over America. We can leave him there in the bush with his porters and his field assistants. His energy is unbearable.

Just as important as his collecting, if not more so, was his microscopy. Technical developments during the nineteenth century were rapid and Eisen kept abreast of them. He was chairman of the San Francisco Miscroscopical Society for a time, and his reputation spread as far as Europe when he invented a kind of ultraviolet-light filter and a different technique for the microscopy of the particularly diffuse details of the inner organs of earthworms.

The microscope opened up a whole new world. His discoveries were many and, in some cases, pioneering. A Californian salamander prompted studies of embryology and the like, which seem to me to anticipate modern stem-cell research. Later he worked on the composition of blood, including human blood, and for a while he was even working on the mystery of cancer, although on that occasion he was wrong. But his discovery of a sort of parasite, very small and resident inside the seminal vesicle of a Guatemalan earthworm, was a real success.

. . .

Once again I am reminded of Strindberg. The summer of 1891 was his last here on this island. He was lonely and miserable, but at that stage he was also profoundly interested in natural history. He was still married to Siri von Essen but only in a formal sense. She had left him and now he needed a microscope.

He had just made the acquaintance of the botanist Bengt Lidforss, and microscopes had come up in conversation between them earlier in the summer. Lidforss had offered to get him one, but Strindberg had obviously organized that for himself. On 19 June he wrote: 'Thanks for the offer of the microscope! But I have one myself now, exactly like yours. Have examined my sperm! Very lively drama of thousands of hot-tempered young Strindbergs which seemed depressed after two hours searching for an egg. Died of unsatisfied sexual drive at 4.00 p.m. at +13° C after having been born at 11.30 a.m.'

He was lonely, as I said earlier. And it was nearly midsummer.

He left the island in August, never to return. He travelled over to Dalarö instead and started painting again. I don't think he ever felt any desire to return. His first letter from Dalarö was no more than a requisition he sent to a cousin in Stockhom. The order was short and to the point: a set of guitar strings, please, and a dozen condoms. The biggest size.

Natural history, no doubt about that – but what's that compared to art?

. . .

Gustaf Eisen's relationship with the fine arts didn't blossom until quite late, and we'll get there by and by. But since Strindberg has once again popped up it's worth taking the chance to fill in a little background. Remember the words uttered by the narrator in the short story 'The Recluse' when he entered the room: 'a fine library, a valuable microscope and an easel'.

Eisen's interest in painting was of long standing, and when he was a bedridden child in Stockholm he had actually earned money by colouring plates in books. Later, in Visby, he came under the influence of Johan Kahl. Kahl was the fifth teacher and the one I avoided naming earlier for fear that my narrative would be burdened with an excess of pedantic detail. Kahl was a watercolourist, closely connected to Palm, Scholander and Gellerstedt and thus a well-known name to anyone interested in Gotland landscape painting. He was also a teacher of both Latin and drawing at the school, and Eisen had an aptitude for both. In fact, before zoology took over completely, Eisen had contemplated art as a possible career. Late in life he reported that painting watercolours had been his meal ticket in his very early days in California, though none of his paintings from that time are known.

The explanation for the easel in the room in Uppsala was that, alongside other subjects, he was being taught painting by the university drawing master Johan

Wilhelm Carl Way (1792–1873), the well-known painter of glass and miniatures.

Strindberg may be feckless in many ways but in this case he can provide us with some solid information. Thanks to the publication of *The Red Room* he was already famous by the time he was thirty and he died relatively young, at the age of sixty-three. Consequently most of his correspondence was preserved and pretty soon being thoroughly analysed. Virtually everything he had anything to do with has been the subject of research, which in turn has resulted in weighty volumes. That is as true of his painting as of anything else. It's said that there are connoisseurs on the Continent who rate Strindberg higher as a painter than as an author. I'll stay out of that discussion, but there can be no doubt that some of his paintings seize your attention. *Wonderland* is one such. I can't get it out of my mind. *Night of Jealousy* is another. He painted it in Berlin for his future wife, and it was stolen from the Strindberg Museum one February day in 2006. It's back there now, the police having found it by chance behind a chest of drawers when raiding a druggy's house in Vallentuna.

Strindberg's interest in natural history continued in the 1890s, though he also found time to marry his second wife, Frida Uhl. When he wrote to Eisen from his new home in Austria in the summer of 1894 the tone is hearty:

Gustaf Eisen, Brother!

I'm sending you herewith a book, the fruit of my labours over the last ten years. It was from you – in the churchyard in Uppsala in 1870 – that I first heard of Darwinism, the transformation. Here you will find it applied to chemistry for the first time. It's been greeted in Sweden as a manifestation of madness and people bewail my fate! Let me know your opinion, even if it's only on a postcard.

Happened to see in a newspaper that you were in San Francisco and are a professor. Of what? Botany? I am now married for the second time and recently had a child by my second wife. Peace and greetings.

August Strindberg

The book he sent was *Antibarbarus*, and I feel certain that Eisen would have taken the time to read it even though he must have been extremely busy at that time, working on the discoveries made during the many expeditions he was taking part in. More or less every year he spent long periods travelling to collect specimens, particularly down in Baja California and other parts of Mexico. And it was also during the 1890s that he wrote his magnum opus, which dealt with the earthworms of the whole of the West Coast of America. After his years as a gardener, this work marked his grand return as a zoologist; it is the culmination of his scientific career and is still used. It's almost impossible to get hold of a copy: the illustrations alone – magnificent colour lithographs that Eisen had made himself – explain why.

He probably recognized that Strindberg did not have what it took to be a good naturalist, but his response must have been reasonably positive because he received another letter within a couple of months, this time from Paris. It was a cry from a friend in need:

Dear Gustaf Eisen,

Sincere thanks for your letter. You can't begin to realize the relief it brought. But, as the author of Antibarbarus, *I am still the idiot and 'producer of humbug and nonsense' as far as Sweden is concerned. Couldn't you arrange for some of your chemists to provide a short and decisive statement that they consider the book 'sound and not in the least crazy'? Or better still a newspaper review that you arrange to be placed in* Allehanda, Aftonbladet, Dagbladet *or* Svenska Dagbladet*? There are still ten parts of the work in manuscript form and I can't get them printed. Why can't I be allowed to experience success instead of having to lead a miserable existence, scorned, mocked, starving and ineffective? One word from you on paper could change all that!*

Yours
August Strindberg

I survive by painting pictures – something else that you taught me!

That final addition is interesting. It is well known that Strindberg did a great deal of painting during this period and *Wonderland* – now in the National Museum in Stockholm – is one of those paintings. The question is, what did Eisen have to do with it?

As usual I had a stroke of luck, at least in the sense that the question arose at a time when I was becoming bored with Eisen, as tended to happen at regular intervals. It meant I could devote myself to Strindberg's painting instead, and when I happened to stumble across a previously unknown little painting things became really exciting. It was a small unsigned piece, hardly bigger than a postcard, depicting a navigation marker in the foreground, a ship on the horizon and some birds. The moment I saw it I was reminded of a passage in Strindberg's autobiographical novel *The Son of a Servant*: 'Johan was forever painting the sea, with the coast in the foreground, enormous pine trees, bare skerries a little further out, a whitewashed beacon tower, a navigation marker. The sky is usually overcast, with light – weak or strong – breaking through on the horizon; sunsets or moonlight, never clear daylight.'

My first thought was 'Take a look at that! A Strindberg!'

The art market shuns the light. That's part of its charm. There is often no mention of where valuable paintings come from; they just emerge and change owners. Knowing looks are exchanged, no questions are asked, and even if my Strindberg – which is what I tend to call it – is neither particularly valuable nor in possession of a dramatic history worth concealing, I am still reluctant to say how I came across it. I need to finish my investigations first. I'm not in any hurry.

The trouble with the painting is that it is likely to have been painted by someone else. It's a bit too good to be by Strindberg – unfortunately. There were navigation markers all over Swedish waters in those days, and good-quality marine painting was by no means restricted to dedicated artists: it was also practised by romantically inclined sea captains, pilots and many others who had the sea in sight day after day. But we should always rely on our gut feelings, so I set about doing some preliminary research anyway and, as it turned out, that opened up a stimulating and exciting prospect.

I started with the panel itself. The painting is on thin card, which, when I first acquired it, was pasted rather carelessly on to a wooden board: mahogany, I think, and as thin as the card. I wondered if there was anything written on the back of the card, and, although I realized it might reveal unambiguous proof that Strindberg had had nothing to do with the painting, I couldn't avoid looking. It wasn't difficult to separate the card from the board and then, all of a sudden, my painting more or less fell into place in terms of both time and space.

The card turned out to have come from some kind of packaging, a folder or a carton, which carried a printed advertisement for the court photographer Johannes Jaeger. Jaeger was in business in Stockholm until 1890. This doesn't tell us much, of course, since Jaeger's business was a big one and his adverts widespread, but the discovery nevertheless caught my imagination and kept me busy for many days. There was a connection, however vague. 'Herr Jaeger's catalogue contains over a thousand photographs, initially exclusively of Swedish artists but then also of various works of art in the National Museum . . .' That is what Strindberg wrote in *Dagens Nyheter* in 1874 during his short career as a journalist. The headline announces 'Cut-Price Versions of Art Works' and the article deals with the mass-produced reproductions that Jaeger was printing at the time. That doesn't prove anything at all, though I did learn a good deal about Johannes Jaeger, who was born in Berlin in 1832.

Then, as usual, I studied the painting under the microscope and that was when I came up with an even more elegant way of showing who had wielded the brush. I have an excellent microscope, so good that you can see things that hardly exist, and so I discovered two things. First of all, a pig's bristle, which gave me an idea, but before I'd managed to think it through I caught sight of something even better. A hair! And better still, if I'm not mistaken, the hair was an eyelash, a human eyelash, root and all, stuck in the varnish. At which point I considered the problem to all intents and purposes solved.

Many years ago I was commissioned by the Academy of Science to write a book on genetic archaeology, and in the process I made the acquaintance of a man, a brilliant scientist, whose exciting work at a Max Planck Institute has elevated the whole discipline to new levels. The information he has managed to reveal by drilling the molars of 30,000-year-old Neanderthals gives us a good indication of what he might be able to extract from this eyelash, or from the pig's bristle.

All I needed was either Strindberg's DNA, which shouldn't be impossible to access, or a bristle from the same pig – that is, the same brush – stuck in the varnish of any one of his better known paintings. The thing is, he was so poor at that point that he wouldn't have changed brushes until they were completely worn out. Elementary! Or at least no more difficult than trapping a car thief, as had recently occurred in Finland, by performing a DNA analysis on the stomach contents of an engorged mosquito that was buzzing around in the abandoned vehicle.

To be honest, I get the impression that connoisseurs of art lag behind somewhat in this area. They seem to be satisfied with basing their judgements on style and brushwork, and are then happy to devote their careers to defaming one another and engaging in open warfare about which of them has come up with the right guess.

And then there is the issue of a name: give a painting a name and it will become known – I'm contemplating *Earthquake Birds*. The rest will happen of its own accord. Strindberg himself once described what happens: 'The painting in itself was not perhaps of any great value, but once generations have stood and admired it, the embryo is fertilized, the egg hatched. Time and people have provided it with its patina.' How true, how true!

I had now recovered to the extent that I could return to Eisen and take an interest in his presumed involvement with Strindberg's career as a pictorial artist. The answer I was seeking was to be found in *The Son of a Servant*. Let's return to Uppsala for a moment.

> *When he was most depressed he would go to visit his naturalist friend. Seeing his herbaria and microscope, his aquaria and his physiological slides, cheered him up.*

Most of all, however, it was the quiet peaceful atheist himself, a man who let the world take its everyday course because he knew that in his own modest way he was working for the future more than the poet with his convulsive and spasmodic progress. But his friend also had an aesthetic side to him – he painted in oils – and Johan found this fascinating. Imagine having the ability to create a green landscape, which can then be hung on the wall, in the middle of the mists of this dreadful late winter!

'Is painting difficult?' he asked.

'Heavens above, no! It's easier than drawing. Give it a try.'

Johan, who had already been brave enough to compose a song with guitar accompaniment, didn't think painting would be completely impossible. He borrowed an easel, paints and brushes and went home and locked himself in.

Chapter 15

The Case of Esaias Henschen

Carolina Rediviva, that inexhaustible library just a stone's throw from the castle in Uppsala, is home to four large cardboard boxes that contain those of Gustaf Eisen's posthumous papers that are still more or less unsorted. They hold a jumble of documents and old photographs, of which it was difficult to get a complete picture. It was while mining the contents of these boxes that I happened to come across a dog-eared leather-bound notebook labelled 'Mysterious Matters'.

I had known for some time that Eisen took an interest in occultism. He had written about it in a melancholy letter to his friend Stuxberg as early as the autumn of 1879. Six years had passed since their paths diverged, and for the last four of those Eisen had been toiling in the Fresno vineyard. He had worked from early morning to late evening as the boss of eighteen Chinese workers and almost as many white, in spite of which he had not earned any money to speak of because the vineyard was not yet productive. The thought of returning to Sweden and making a career as a natural scientist seemed an ever more distant one – after all, by now he hardly knew anyone who could help.

Life in California, moreover, had made him a different person. 'I have changed considerably during the past six years and it's possible that you wouldn't enjoy my company now.' Theosophy and occult sciences, he continues, have opened a whole new world 'more wonderful than the one the bucket dredges up from the depths of the sea or is revealed by the microscope'.

Eisen's new idol and companion was Madame Blavatsky. In his opinion her book *Isis Unveiled*, published a couple of years earlier, was 'one of the most remarkable works of the present day', and I get the impression that its attraction was somehow linked with all the hardships of life as a pioneer. No one gets anything for nothing, most people fail, many go under. The same letter told of a shared childhood friend who had visited, destitute and sick. Eisen buried him in the hard soil.

Many years later he would refer to these first years in Fresno as 'my Babylonian captivity'. He could, of course, always dig for earthworms and catch centipedes in the mountains – as a comfort, to help him endure – but clearly something more was needed, something of a more spiritual order.

Helena Blavatsky (1831–91) was born in the Ukraine and lived an adventurous life, much of which has to be seen as invented, primarily by her. What is incontrovertible, however, is that she founded the Theosophical Society in New York in 1875, and Eisen, as we shall see, was an early adherent and more than just a passive admirer.

What seems more or less certain is that Blavatsky grew up in an eccentric part of the Russian aristocracy.

Her mother was an authoress who died young, after which the girl was raised by her maternal grandmother, who is recorded as having been an archaeologist and botanist with a large library. The grandmother, moreover, was married to the governor of Saratov on the Volga and seems to have been indecently rich, with all that implies in terms of luxury and servants to act on one's merest whim.

In spite of that – or perhaps because of it – Helena stood out as rather strange even as a child. It's said, for instance, that she could communicate with animals; even today there is nothing very remarkable about that, perhaps, except that in her case the gift included an ability to communicate with stuffed animals. That has to be seen as taking things a bit far. I, too, have my sad days when I unburden myself on the hundred-year-old peacock that stands looking haughty in the corner of my study, but I certainly don't get a response – a fact for which I am more than grateful.

Anyway, Madame Blavatsky, as she came to be called, acquired her surname as a teenager when she married a count more than twice her age, a man she ran away from very soon, going off to Constantinople and then out into the wider world. What followed is, as I've said, unclear: her adventures in Tibet and elsewhere during her years of travel, lasting roughly from 1849 to 1873, are colourful but only credible to the gullible. Somehow or other, however, she must have boned up on spiritism, which was a flourishing industry at the time, and that is where her gifts came into their own.

Seances led by a medium, usually a woman, were in the process of becoming a very popular form of social life, particularly in America. Laconic messages from deceased relatives and friends, even if only in the form of knocking, seem to have exerted the same pleasing fascination as the empty phrases that currently abound in the social networks of cyberspace, although these days it tends to be contact with living people that is sought.

The demand was insatiable. At the end of the nineteenth century there were thousands of mediums in California alone, and even though the seance itself, given the mild eroticism of gathering in darkened rooms, was more rewarding than all the supernatural phenomena the mediums might conjure up for the suggestible, the stream of followers can also partly be explained by the breadth of the doctrines of theosophy. It claimed to embrace virtually everything. Natural science, with Darwin at its head, had undermined the authority of the Church; ancient creation myths were jettisoned; similarly primitive gimmicks like the promise of Paradise, and even the threat of the terrors of Hell no longer worked. But all this had failed to diminish the human yearning for mystical experience.

The theosophists wanted to put absolutely everything in one single bag of tricks, and they set about the task with a vengeance. The theory of evolution, transmigration, astronomy, alchemy, Christian and Eastern mysticism, meditation, astrology, mathematics, angels, fairies and everything else right down to the daily weather forecast were stirred into a spicy broth, which was then cooked according to a secret recipe handed down from a secret brotherhood called the Masters of the Universe. Ordinary stuff and nonsense, then, not unlike its corny latter-day equivalent, New Age. There is an interesting difference, however, in that these days the hankering for spiritual rebirth, clairvoyance and other obscure esoterica are mainly attractive to marginalized and less educated groups whereas at that time they were socially acceptable in much wider circles.

People desirous of power and status today, whether in scientific, political or cultural spheres, have to rely on the old religions of the book if they really want to live out their faith in the irrational. Ghosts are not sufficient. Eisen's age was more permissive in that respect. Strindberg, for instance, was certainly regarded as crazy, but if any modern author of his rank had written *Inferno* the judgement would have been much harsher. He, too, came under the spell of Helena Blavatsky, who was a guru for many people at the time. One of those closest to her explained the attraction: 'It was her eyes that attracted me. They were the eyes of someone I felt I must have known during lives lived long ago.'

Madame Blavatsky later moved to India, where she started the journal *The Theosophist*. That was in Bombay in 1879, and the very next year the journal contained an article – 'A True Dream' – by Gustaf Eisen. It's a strange story from his childhood in Visby, but without any real edge. It's really of interest only as evidence of his contacts, not otherwise. The boxes in Uppsala, however, had much more than that to offer.

. . .

I opened the black folder and started to read. It proved to be a collection of shortish stories, written while Eisen was living in New York, that is to say, after the First World War. The subjects varied, but what the narratives had in common were the occult phenomena he had encountered on his path through life. His intention may have been to publish them in *The Theosophist* or in one or other of the esoteric periodicals he regularly contributed to – *Lucifer*, *The Progressive Thinker*, *Banner of Life* and all the rest. The boxes in Uppsala contain a wide range of newspaper cuttings of that sort.

One neatly typewritten story in the folder – 'The Man Above Me' – takes place in 1916, when Eisen was renting a room on 156th Street in the Bronx and was in the habit of sleeping with his window open, partly because the summer was very hot that year and partly to remain on friendly terms with a large stray cat that used to climb in at night and sleep in his room. One night he was suddenly woken by a pistol shot, after which a series of inexplicable things happened. On the whole the story is not much of a success.

I browsed on, while the librarian in the reading room kept a careful eye on what I was up to.

'The Gnome' is an account of Eisen's sighting of a gnome in his youth. It was on Gotland. Nothing out of the ordinary, since everyone knows there were plenty of gnomes around in those days. 'The Libretto', written in March 1929, is a short and vaguely mysterious anecdote related to the long-lost opera libretto Eisen wrote in 1899. Nothing very special: the story of the opera is fantasy, of course, and we shall have reason to return to it.

A collection of rubbish, I thought. The kind of thing we should spare our surviving relatives the trouble of sorting through, but I carried on anyway because you can never be sure with unsorted boxes and, true enough, I came at last upon a story that led me onwards. Just a couple of pages written in March 1921: 'The Case of Esaias Henschen'.

The story began in Fresno in 1879, during the difficult years when Eisen found recreation and excitement by visiting Mrs Butler, a medium who lived in the neighbourhood and whose speciality was automatic writing. She coaxed messages from the spirit world by placing her hand over a small piece of chalk on a slate. Eisen took it into his head to get in contact with an old friend:

But before I go on I should mention that during my time in Uppsala I was a lodger in the house where Judge Henschen lived with his family. It was his house. I paid 160 kronor a year for two unfurnished rooms. Anton Stuxberg, who later became an explorer and the head of the museum in Gothenburg, lived in the same house.

Henschen had three sons. Esaias, the middle son, was in poor health and suffered from tuberculosis, for which reason he spent a lot of time out in the fresh air. He had business with some Swedish emigrants who had founded a settlement in Florida – he had transported them there from Sweden. He travelled back and forth every other month and we thought he was mad to travel to America and back six times a year.

When he was about to set off on his last voyage, Stuxberg and I scraped together a hundred kronor and gave them to him in exchange for a promise that he would use the money to collect the insects and other creatures which we assumed Florida to be crawling with. I seem to remember that this was in 1871, a year or so before I myself travelled to California. At any event, the promised natural history collections did not arrive during the time Stuxberg and I were in Uppsala and we assumed that Esaias had forgotten all about it or that he had not managed to find anything to collect – he was, after all, not a zoologist.

That's the start and before long the chalk can be heard rasping on the slate. The message from the other side is quite clear. What is written on the slate says: 'Forgive me for letting you and Stuxberg down, but I died before I could carry out your wishes. I am grateful for the honour of being permitted to communicate this. Esaias H—n.'

Aha, I thought, yet another failure. But I have to admit that my curiosity made me read the rest of the story, because, in fact, the name Henschen wasn't unknown to me. Moreover, the author made a striking chronological leap forward to 1904, which was when the story had an unexpected resolution. A resolution that also revealed the author's doubts about his theosophical beliefs.

I went to Sweden that year and during my stay my friend Georg Törnquist, the famous actor, and I decided to go to Uppsala to revisit the places where we had enjoyed such good times in our youth and where we had even done some work. We were there for three days, if I remember rightly, and the newspapers reported our visit, which seemed to arouse public interest. We went to Henschen's old house and found it unchanged, but we could not go inside since there are strangers living there now.

We then returned to Stockholm and the following day I had the surprise of my life. Georg Törnquist received a letter addressed to me. It was from Esaias Henschen, now the manager of a bank in Uppsala. He had read in the paper that I was there, had tried to get in touch but was too late. He enclosed a cheque for a hundred kronor, with many years' interest, and he wrote that the parcel of interesting animals he had sent from Florida thirty years before had got lost on the voyage. He had been really concerned that it had taken until now for me to get my money back! An honest man! How few they are!

Stories of that kind tend to transport me into a state of ecstasy. A clue! A message from the other side! Esaias, now I've got you! I ran – literally – to the section containing biographical reference works, where I soon discovered that Esaias Henschen (1845–1927) was the older brother of the brain surgeon Professor Salomon Henschen, who is famous even now, partly for performing the autopsy on Lenin and partly for being the progenitor of a string of outstanding medical men, artists and authors.

Esaias seems to have lived a quieter life, and to my great disappointment I could find little written about him. He was just one bank manager among many. But a couple of days later I got a bite.

He had actually been in Florida in his youth, and, thanks to an old emigrant association over there that still preserves his name, I discovered that – unlike his brother Salomon – he had been the forefather of ordinary mortals rather than celebrities. But there was one of them who had made a name for herself, a great-grandchild, a Swedish poet, famous even in Florida, whose name is . . . No, this is not the place to reveal it!

Nothing surprises me any more. I gave up on that long ago. As a collector of flies I quickly learned to stand motionless and wait, net at the ready, for weeks if need be. The real rarities will come sooner or later, believe me. The poet's father was like that, a warm-hearted man, an old major general who gave me a thorough account of his grandfather's long and colourful life. None of which is relevant in this context other than as an example of the motivation that drives a collector. Flies or stories, it doesn't matter which.

. . .

We ended up in Monterey, up the coast. From the mountains of the Sequoia National Park we travelled straight across the valley as quickly as we could, down to Cambria on the Pacific coast. Two worlds as different as chalk and cheese in the same day. Then we drove north along Highway 1 towards San Francisco, a road that snaked high above the shore, the wild greenery of the coastal mountains forming a wall on one side and the ocean forming the horizon on the other. But before we reached Monterey, where we stayed in Pacific Grove for a few days, we visited Hearst Castle. It lies close to the road.

The press baron William Randolph Hearst, the model for the protagonist in Orson Wells's film *Citizen Kane*, owned sixty-two square miles of land here and he built a fairy-tale castle that now stands on its mountaintop like a monument to the madness of collecting. Hearst threw fantastic sums of money into his manic search for art and antiquities, and, since his interests and tastes changed a great deal during the almost three decades it took to build the castle, the whole thing is almost comic in its variety.

Nothing was impossible. If Hearst wanted a Roman temple he bought one, had it dismantled and shipped over to California, where the architect Julia Morgan did whatever she could to fit it into the construction. He would buy whole rooms from Renaissance palaces in Europe, complete with panelling, ceilings, windows, furniture, floors and carpets. Everything is over the top: medieval armour, sculptures, books, silver, mosaics and tapestries. Particularly tapestries – enormous tapestries and huge rooms to hang them in.

He had inherited his interest in old textiles from his mother, Phoebe Apperson Hearst (1842–1919), who was also incredibly rich and happy to collect anything that took her fancy, which was pretty well everything. In her defence we should say that private collecting often bordered on philanthropy. Her collections ended up in museums, and since she hired artists and the like to hunt through the antiques markets she did her bit for the employment of many people.

It's Phoebe Hearst we can thank for the book *Maya Textiles of Guatemala: The Gustavus A. Eisen Collection*, which wasn't actually published until 1993. And for the collection, too, of course – the world's finest and most complete collection of antique Maya textiles.

Around the turn of the century Phoebe Hearst recognized Eisen's talent. He had just published a book on figs, a successor to his book on raisins, and at the same time had ceased being curator at the California Academy and started a photographic studio in San Francisco instead. He was a gifted photographer, took very fine portraits, and for several years he won high rankings in various national competitions for art photography.

It couldn't have been a difficult decision for Phoebe Hearst once she had decided to collect Central American textiles. Eisen was an old and experienced Guatemala hand, and he was both a great collector and photographer. What could be better? She sent him off for about a year – that was 1902 – and the collection he returned with must have been kept somewhere that escaped the earthquake. There can be no doubt that Phoebe Hearst was pleased with the way he fulfilled his commission, and there is good evidence to suggest that she kept Eisen afloat in economic terms right up to the First World War.

I think they got on well together, even in a personal sense. Like Eisen, Phoebe Hearst was a freethinker of the New Spiritual variety. At about the same time as they became acquainted she converted to Bahá'i, a religion founded in Persia in 1844, the leading prophet of which was called Bahá'u'lláh. The doctrines of Bahá'i are essentially that all the great religions and holy books are more or less equally good and valid, but that Bahá'u'lláh's own holy book, *Kitáb-i-Aqdas*, has a slight edge on them. A splendid religion by all accounts. There are quite a few million Bahá'i believers these days, some even in Sweden, where early congregations were established in Hultsfred and Jokkmokk.

. . .

Now, however, we had arrived in Monterey, south of San Francisco, where we checked into a hotel by the sea in Pacific Grove: it had a balcony on which I spent several days sitting attempting to learn the names of birds I had never seen before – surfbird, Heermann's gull, Brandt's cormorant – while the sea otters carried on fishing unconcernedly for mussels just offshore.

Eisen was a jigsaw puzzle with too many pieces. Why did he return to Sweden? He went back to his homeland twice, once in 1904 and again in 1906, but there are no straightforward answers to what lay behind these trips. In a purely private sense he was in a time of transition and he always had several irons in the fire. There is, however, one explanation that is better than the others. It emerged in Monterey when we were out wandering along Cannery Row in the heart of Steinbeck country, where the old canning factories have now been converted into an enormous marine aquarium.

During the years before the earthquake there had already been plans for something of that kind, a combination of tourist attraction and marine-research laboratory. As usual, it was one of those eccentric millionaires who had decided to finance the project. It was to be located in Golden Gate Park in San Francisco, and Eisen had been invited to be its managing director. He was also to be responsible for drawing up the guidelines for the construction, and consequently he needed to travel to Europe to study aquaria and marine-research stations. Once there he took the opportunity to visit his old friends in Sweden. Stuxberg was dead by then, but Strindberg was sitting like an old owl in his flat on Karlavägen in Stockholm.

In his *Occult Diary* for 17 April 1904 Strindberg wrote: '... night. Dreamed about Berzelius very vividly, also about Gustaf Eisen.' And in a cryptic note in the September of the same year we can read about dinner at a Lidingöbro restaurant at which Eisen succeeded in convincing Strindberg to accompany him to San Francisco to spend a couple of years learning English and perhaps starting something new. But Strindberg backed out at the last minute. There is a later addition in the margin: 'cf. April 1906 San Francisco burned.'

There was always some sort of hitch! Plans for the great aquarium were shelved following the earthquake. It would be very easy to get the impression that Gustaf Eisen's long life was characterized by recurring failure.

On the other hand, however, I think he quite enjoyed it and, when all is said and done, what would human life be like if there weren't any snags?

Chapter 16

Legends of the Holy Grail

There are many unpublished manuscripts among Eisen's posthumous papers. The volume on the history of glass beads, along with 40,000 watercolours, ended up with the Royal Swedish Academy of Letters, History and Antiquities in Stockholm. There is also a large-scale libretto for an opera composed by Carlos Troyer (1837–1920), with whom Eisen was collaborating around the turn of the century; the sheaf of papers is moth-eaten and difficult to read but we can tell that the story takes place somewhere in the south-western states of the USA during the prehistoric period. Troyer was very interested in the original inhabitants of the country and their music. The opera was never performed.

There are great bundles of historical and literary manuscripts of all sorts in both Stockholm and Uppsala, and I doubt that anyone other than the author himself has read all of them. There must be thousands of pages, mainly written during the first decades of the twentieth century. The biggest manuscript is called *The Alhambra Tales*, a collection of more or less fantasy stories with historical settings, possibly influenced by Washington Irving. The rejection slips from big New York publishing houses serve to confirm the impression that Eisen's literary talents resembled Strindberg's ability as a chemist: he was a quack, a dabbler.

Perhaps he wrote in order to pass the time. He moved back to San Francisco after the earthquake and helped with the reconstruction of the museum, but he was soon on the move again. Between 1910 and 1915 he was roaming around Europe, sometimes in the company of the painter and collector of antiquities Carl Oscar Borg, who also depended on the economic support of Phoebe Hearst. It's not clear whether Hearst's patronage was linked to anything in return, but these were the years when Eisen concentrated mostly on research into beads and antique glass. He lived alone in Rome for a long time, writing the whole time.

Or was he perhaps writing to earn some money? I have found a film script – *The Stolen Combination* – among his papers, and several short stories that more or less fall within the genre of science fiction. One of them with the title 'Ultraviolet and Infrared' is rather reminiscent of H. G. Wells's classic *The Invisible Man*, and it may well have been precisely this lack of originality that lay behind the reluctance of publishers to publish what he wrote. The outbreak of war certainly wouldn't have made the situation any easier.

It is against that kind of backdrop that we should see Gustaf Eisen when he arrived in New York, sixty-seven years old and probably rather dejected, in April 1915. For some reason or other Phoebe Hearst was no longer providing support, the world war had closed off Europe, he still had his contacts with the fig experts in the Department of Agriculture in Washington but that hardly provided an income, and he had left zoology for good. His intention was to stay for no more than a couple of days and then take the train home to San Francisco, perhaps to start a new photography studio. But that is not what happened.

It all came about very quickly. It started with him walking along Fifth Avenue in Manhattan and seeing in the window of an antique shop a necklace of antique glass beads of a kind he had never seen before, which in itself was quite amazing. He went in, introduced himself as an expert and asked permission to do a painting of the beads. The owner of the shop, a young Syrian from Paris by the name of Fahim Kouchakji, had no objection. Quite the opposite, in fact. He seems to have recognized immediately that Eisen was a man who could do great things so, once the beads had been documented on the watercolour pad, he offered to show the Swedish stranger one of his latest acquisitions. A silver chalice from Antioch.

A truly remarkable piece, discovered a few years earlier. A chalice, plain on the inside but decorated on the outside with animals, plants and a dozen figures who seemed to be at table. Eisen had scoured all of Europe's art museums with the same thoroughness as he once searched the jungles of Guatemala so he could say with certainty that there wasn't another one like it. But he had seen several smaller chalices of the same shape and proportions in the Louvre – they were from the time of the Emperor Augustus.

Fahim Kouchakji, who was a very wealthy man, asked Eisen whether he would consider writing a report on the chalice. Just a couple of pages. 'I accepted the task,' Eisen comments much later, 'although I could see it would take several days, perhaps as much as a week.'

It took him eight years. When the book eventually appeared in 1923 – *The Great Chalice of Antioch* – it was in two magnificent volumes, weighing fifteen pounds, printed on handmade paper and with top-quality illustrations of the most minute details.

It's impossible to say what his motivation was or what he believed. Kouchakji wanted to make money, no doubt about that, but Eisen? What drove him?

I don't believe he was ever completely convinced

that the chalice was the Holy Grail or that the people portrayed on its outer surface were Jesus and his disciples. But I may be wrong. A letter to Svante Arrhenius, who is probably best known now as the discoverer of the greenhouse effect, suggests the opposite: 'This object is not alone [*sic*] the earliest Christian work known, but it is the most important work of the kind known, and the most important object of art in the whole world today.' I found this letter, written in English and dated 1918, at the Academy of Sciences in Stockholm.

At any event, I am pretty certain that the chalice suited his disposition and talents as much as Gotska Sandön had done. A good deal smaller than an island, of course, but size is irrelevant: sharp eyes and a good imagination can find all kinds of things even on islands that can be held in the palm of the hand.

The book sets out to prove the age of the chalice and to show that the figures really do depict Jesus and his disciples. Furthermore, by following long chains of circumstantial evidence it asserts that the portraits were probably created by someone who had actually met the thirteen participants at the Last Supper. Plenty of believers all over the world were prepared to swallow the story.

The chalice quickly became famous. When it was put on public display police guards had to be posted because it became a magnet for all sorts of fanatics; and meanwhile, in the background, Fahim Kouchakji was negotiating with speculators as rich as Croesus. The Vatican itself became involved when the mania was at its height. Professional historians, however, were doubtful from the start and the debates were sharp. Nowadays the chalice is considered to be sixth century, not from Antioch, and an oil lamp, not a chalice.

But it was already famous as the Holy Grail and those who believed believed. Full stop. Eventually it was bought – expensively – by the Metropolitan Museum of Art and as recently as 1955 served as model for the chalice in Paul Newman's first film, *The Silver Chalice*. In 1975 *Monty Python and the Holy Grail* also included references to Eisen's old silver chalice.

For many ingenuous people, however, it seems likely that it will always remain sacred, irrespective of what experts and comedians may say. In 1936 the journalist and Eisen admirer Magda Måneskköld said it all – in a single sentence!

> *Before concluding this account we should, in however summary a form, provide a survey of Dr Eisen's later activity – undertaken after he had reached an age when most people seek peace and quiet – in order to see how it was introduced to the sceptical eyes of the world, becoming the object of many years of hostile debate, protest and angry pronouncements from doubting colleagues both on the European and on the American scene; and then to observe how finally – thanks to his knowledge, superior intelligence and unfailing self-confidence – he calmly and comprehensively silenced his critics and defeated his keenest opponents so that they allied themselves with his most loyal adherents in the greatest discovery in the history of our age – that is to say, the discovery of the sacred communion chalice of Antioch.*

Eisen's book is difficult to get hold of, and, since there isn't even a copy in the possession of the Royal Library in Stockholm, I had to summon up all my skills as a hunter to eventually locate one in an antiquarian bookshop in London. As I've already said, it's the most expensive book I possess, as well as being the biggest. Religious humbug, but in a deluxe edition.

It was, however, worth the money. I was particularly taken by the so-called Inhalation Theory. This is the final link in the chain of evidence that – according to Eisen – proves that the chalice is from the right period and authentic. Put in its simplest form, the theory states that up until the first century A.D. Greek artists and their imitators depicted people breathing in, whereas all later artists invariably depict people breathing out. An expert, Eisen claims, can see this and utilize it as a method of dating. We don't need to concern ourselves here with how he managed to conclude that the squiggles on the chalice are actually breathing in – to understand that demands a level of faith that I find impossible to muster!

To my joy, however, I found a completely different story tucked away in this theory, so to speak. In his chapter on the Inhalation Theory, Eisen refers to his good friend, the artist Arthur B. Davies (1862–1928). He was the real expert on Greek inhalation. A mystic. I think we need to work backwards here.

. . .

The circumstances surrounding Arthur Bowen Davies's death in Florence in 1928 were mysterious to say the least, and the whole business was made worse by the fact that news of his demise did not reach the United States until seven weeks later. This was remarkable, particularly given that Davies was one of the most admired artists in the country at that point and many considered him to be the leading figure. Years later some newspapers were still expressing doubts as to whether he was actually dead or whether, perhaps, he had just chosen to disappear.

He had done so before, though no one knew why. The answer would eventually turn out to be a simple one: women. His story is not his own, it is that of his women. Even his art was theirs. Literally.

Davies came from Utica in New York State and studied at the art college in Chicago. After some peripatetic years, during which he made his living as a draughtsman for the Santa Fe Railway in Arizona and New Mexico, he settled down on a farm on the Hudson River twenty-five miles north of New York. Settled down is perhaps a less than accurate way of putting it, but this farm became his address anyway – or one of them – as long as he lived and Virginia Meriwether, the owner of the farm, became his wife.

Virginia came from the South. She was a determined woman who as a teenager had married a drinker who, shortly after they had got married, was stupid enough to point a pistol at his young wife. Which he regretted, because she, too, happened to be armed – and suddenly she was a widow. It was self-defence, of course, and thus an uncomplicated case in legal terms, but the press grumbled all the same and made the girl's life uncomfortable, so she moved to New York and trained as a doctor.

Shooting one's husband was not really very remarkable, but a woman becoming a doctor in the 1880s was quite exceptional. Virginia was one of the first.

Now, however, she and Davies were going to become farmers, a romantic project that rapidly came to nothing, since the latter soon recognized that family life in the country was not what he was cut out for. Virginia was five months pregnant – they had a number of children – when Davies moved down to Manhattan to pick up his career as an artist. He visited the family at weekends, if at all, and since it was sometime before his paintings began to sell his wife had to continue working as a doctor and midwife out in the countryside. She is supposed to have helped 6,000 children come into the world and with the passing of the years she also began to make a decent profit from the cultivation of strawberries.

Arthur B. Davies was a painter in the spirit of the age, profoundly influenced both by the English Pre-Raphaelites and quite soon by such Symbolists as Arnold Böcklin, Pierre Puvis de Chavannes and Edvard Munch. He was also interested in the art of the Ancients, particularly that of the Greeks. Combining all this with unforgettable memories from the Sierra Nevada and other wild parts of the West, he concocted a very individual style, dominated by dreamlike, mystical landscapes of mountains and plains with people in the foreground, usually fine-limbed women more or less undressed, who were performing strange and hesitant dance-like movements.

Davies's romantic melancholy is rather reminiscent of David Wallin, who is not remembered by many people these days, though he was constantly painting his wife, who was extremely beautiful. Against hard competition from Isaac Grünewald, Bruno Liljefors and others at the 1932 Los Angeles Olympic Games, Wallin won the gold medal for painting: that's a competitive field that is unfortunately no longer included.

Davies, however, unlike Wallin, rarely painted his wife. He paid models, younger beauties, and, as we shall see in a moment, the whole of his production from the turn of the century on reinforces the suspicion that male artists' interest in the female body is actually all about sex. There are, I know, other explanations, each more respectable than the last, but it is necessary to fall back on a good deal of naivety and art-historical flummery to draw a veil over sexual attraction, both that of the artist and that of the spectator.

Davies became the highest paid artist of the day in America, and it was notable that women bought his paintings. Artistic life in New York during that period seems to have been an outright matriarchy: Lillie Bliss, Gertrude Vanderbilt Whitney, Abby Aldrich Rockefeller and all the rest of them threw themselves enthusiastically into the art market, and their private collections have since formed the basis of several of the great art museums of modern times. They had inherited or married into the unimaginable fortunes raked in by robber barons in oil and banking, and when the time came to spend their money they created an art market that makes Damien Hirst seem little more than a door-to-door salesman.

These women also financed Arthur B. Davies's greatest triumph, the 1913 International Exhibition of Modern Art, what came to be known as the Armory Show. There has seldom if ever been an art exhibition of greater importance. Things that had been current in Paris and elsewhere in Europe for some time – Fauvism, Cubism and so on – struck the New York public like a bolt of lightning. Ever since then American art history has been divided into two distinct phases – before the Armory Show and after it.

Davies frequently travelled to Europe and was consequently aware of trends in the art world. Modernism was quite simply the future, and he therefore decided to ship over a representative sample of his favourite artists for exhibition in the United States: Picasso, Cézanne, Matisse, Léger, Kandinsky, Braque, the whole lot of them. He managed to get hold of no fewer than eighteen works by Van Gogh and fourteen by Munch. Even their predecessors were well represented: Ingres, Delacroix, Courbet, Manet, Renoir, Monet, Corot, Degas, Whistler and so on.

The result was predictable. A scandal! Art critics condemned the whole business and the president, Theodore Roosevelt, who was being guided around personally by Davies, came to a halt in front of Marcel Duchamp's painting *Nude Descending a Staircase No. 2* (1912) and uttered the famous words 'That's not art' – words that ever since have caused anyone with an instinct for self-preservation to refrain from calling humbug humbug. A good deal of bowing and scraping and outright cowardice in current art criticism can be traced back to this statement. No one is prepared to take the risk of having the wrong opinion.

Well, very quickly everything turned round and became a huge success; the conventions of academic art, that had lain like a wet blanket over American art until then, soon became history. Davies, naturally enough, didn't have much time to be at home now that he was in high favour. Solo exhibition followed solo exhibition, and in 1920 he was shown at the Venice Biennale. And, as everyone knew, he had to be left in peace to create. That's the kind of person he was – rather private.

It proved to have rather less to do with art than with the fact that for a long time he had been secretly living under a false name with one of his earlier models, Edna Potter, in an apartment in Manhattan. He went under the name of David A. Owen, engineer, and they had a daughter, Ronnie. Virginia back at the farm was clearly not a woman to be messed with, so it had been necessary for him to lie low and that had only served to enhance his reputation as a great painter.

Everything was going well until Ronnie was twelve years old and beginning to take an interest in her parents' background. Davies, alias Mr Owen, saw the risk and moved Edna and Ronnie to a flat in Paris. This was in 1924. But there were many Americans in Paris who might recognize him – he was famous by this point – and they soon had to move on once more, this time to Florence.

Davies began to become paranoid, and with good reason. At the time of the 1920 census he had felt sufficiently secure in his double life to register separate identities at two different addresses, but now the game was coming to an end. He'd had a new model for many years and was probably involved in a sexual liaison with her, too, but he managed to keep that one quiet.

He died in Florence. Edna had him cremated and then took the boat to New York, where she visited Virginia on the farm and told her everything. You can imagine the scene: two women, the older one with several children and grandchildren, the younger one accompanied by a furious teenager, Ronnie having only learned of her father's true identity on the boat on the way over.

That wasn't the end of it. Virginia and Edna decided to hold their tongues and hide the scandal. They travelled to Europe together to clear up any traces left behind and to save what could be saved. Davies' private art collection was enormous. There was a fortune in Paris and Florence, and even more in New York when they emptied his studio apartment at the Hotel Chelsea, a place that even now can boast of its unusual guests and tragic endings. That's where Dylan Thomas died of alcohol poisoning and later, in a different room, Sid Vicious of the Sex Pistols killed his girlfriend Nancy Spungen.

. . .

Why is it that hotels that want to boost their image always draw attention to those unfortunate guests who were unable to check out? I know it's a tough business, but all the same. When I was invited to speak in Växjö a year or so ago (the cultural elite of the area had been seized by an unexpected interest in hoverflies) the organizer pointed out that they had booked me into the town hotel and added *en passant*: 'That's where both Birger Sjöberg and Christina Nilsson died'.

. . .

The two women sold off the art, which included sixteen works by Picasso, and came up with a convincing story as to why it had taken seven weeks for news of the death to be announced.

Eisen was probably one of the very few people who knew of Davies's double life. There is no doubt that the two of them got on well: both were collectors and enthusiastic theosophists with a weakness for mystical theories, such as the one about inhalation. And, to be fair, it should be said that Eisen was not completely accurate when he talked about the origin of the inhalation theory: he said it came from Davies, whereas in fact it was Edna who had come up with it.

Edna Potter Owen was a dancer before she became a model. She was profoundly influenced by Isadora Duncan and through her she had become interested in Greek dance. That is where the inhalation theory came from. I recognized that fact one day when rummaging through Eisen's posthumous papers in Uppsala and found a couple of letters to him from Helena Garretson. Letters dealing with dance.

And who was Helena Garretson? It was the pseudonym of none other than Edna Potter Owen. An art historian and researcher into dance in the United States I am in correspondence with has told me that Edna, under the pseudonym Helena Garretson, wrote a book called *Dancing*. It exists in the catalogue of the Library of Congress in Washington – but it has disappeared. My friend the dance researcher has spent years searching for it all over the world and been unable to locate a single copy. Gone. One of the letters to Eisen, written in Paris in 1924, is signed 'Edna P. Owen (Helena Garretson)'. So he must have known. I'll find that book one day.

Among Eisen's papers there is also a very substantial unpublished manuscript with the title *Legends of the Holy Grail*. If I know him as well as I think I do, he will have gathered everything there is to know about the Grail by way of medieval myths and legends, but it would have been unreasonable to expect to fit it all into the book about the Antioch Chalice. Perhaps he was contemplating an even bigger and more general work about the Grail, viewing it in a wider context – not merely as an artefact but more as a dream.

Chapter 17

The Nest-boxes in Central Park

When *The Great Chalice of Antioch* was published, Gustaf Eisen was already an old man of seventy-five and at the stage of life when a man's biography usually begins to thin out. People start takings things easier, enjoying the leisure and the peace and quiet of retirement. Not Eisen: he continued at the same pace to the end.

There were still three big books to come.

He was now an established art historian and that implied a host of possibilities. First of all he unpacked his European notes about antique glass; not the beads – that manuscript remained unpublished – but glass in the broad sense. During his years in Rome he had been into glass, too, and his contacts with the Metropolitan and other museums of art now gave him the chance to complete that project. The financial mogul J. P. Morgan had been a collector of Frankish and Merovingian glass and his collection was in New York, and the widow of William H. Moore, another of the robber barons, possessed an even finer collection.

Eisen was given access to these and other collections. Fahim Kouchakji was still in the picture both as an expert and as a close friend. He was the one who funded the resulting book, once again a magnificent production: *Glass: Its Origin, Chronology, Technic and Classification to the Sixteenth Century* (1927), a richly illustrated two-volume work of almost 800 pages. It was printed in a limited edition of 525 numbered copies and consequently it is difficult to get hold of, all the more so as it's still considered useful for the classification of glass from the time of the Pharaohs onwards.

The principles Eisen used for dating may be a touch out of date now. That doesn't worry me. I am quite content just to own the book, as a talisman and to enjoy the beauty of the author's drawings and photographs, which illustrate his history as well as that of glass. All his journeys, all his toil, all his joy – and, no doubt, the loneliness, too. In so far as I can judge, what people said at the time was true: he was the world expert when it came to dating old glass.

He was eighty years old by this point and it was time to find something new. Now he returned to the portrait painting he had studied with Carl Way in Uppsala at the start of the 1870s and he did so in a manner that I find both rather touching and also in character. It's as if he wanted to bring to bear everything, absolutely everything he had learned over the years: his zoological eye for anatomical and other detail, his persistence as a collector, his interest in portraiture, and above all his finely honed fondness for systematics.

He set about writing what is his most remarkable book and, in my opinion, his best book. It's also the most difficult to find. *Portraits of Washington* (1932) is as rare as a hoverfly of the genus *Callicera* and is virtually impossible to get your hands on. A weighty 1,000 pages in three volumes, printed in a limited edition of 300. Everything about the work is so beautifully crafted that it's obvious no expense was spared: the paper, the photogravure, the typography, the leather binding with its gold tooling, everything. Not a single Swedish library has a copy. But I do.

Portraits of Washington is a study of all the known portraits of the first president of the United States. Well, not all, perhaps, but a great many. There are some owners, as Eisen tells us in the preface, who are unwilling to show their paintings because they are worried that an expert might identify them as cheap copies. You can understand their concern. But he did manage to see more than sufficient works, the products of about 150 artists, in the form of paintings, engravings and sculptures.

Portraits of the president had very quickly become something approaching an industry in the United States, as often happens in young republics. Long ago, when I was criss-crossing northern and eastern Zaire, as the country was then called, I often stayed in villages where the people owned next to nothing. A little group of huts built of twigs in the forest but, with very few exceptions, they all had a portrait of the dictator Mobutu Sese Seko. There are numerous examples of the same thing even now, and I imagine the tradition lasted for quite a long time in the United States. It can be explained in simple socio-biological terms, not that that need bother us here.

Gilbert Stuart (1755–1828) was the most famous painter of George Washington portraits. He was the man who painted the picture that is on the one-dollar bill. He returned to the subject time after time, and was one of the few actually capable of getting the president to sit still for a while. Later on he copied his own portraits, and then other people copied his portraits, and so on ad infinitum. The first volume of Eisen's book is entirely devoted to portraits by Stuart.

The author made no secret of the fact that he used the same methodology that had made him a leading earthworm expert. The trick was to focus on seemingly insignificant details, the kinds of things that escape the attention of laymen. In Eisen's opinion, ordinary art experts often fail to distinguish the genuine from imitative hackwork because they insist on basing their judgements on commonly held opinions about style and visual verisimilitude.

Eisen, on the other hand, argued that one should sidestep the feelings and taste of the spectator and other subjective factors and observe instead such accessories as the president's hairband and cravat, or imperceptible variations and minor details in the furnishings, even the pattern of the carpet the old fellow was standing on. By means of studies of that kind, carried out with exceptional thoroughness, Eisen drew up a table of criteria.

Meaningless, of course, but tell me what isn't?

The work involved gave him great pleasure, that's obvious even from a distance, and it was probably what kept him alive. He also tells us that he had started collecting material about Washington as long ago as 1915, mainly as a form of relaxation, a way of resting from other more stressful tasks. And I'm afraid to say that I rely on activities of the same kind to keep me going.

. . .

There was something about Eisen that caused me to remember my childhood. I didn't think about it much at first, but later, when I realized that he had never actually got beyond Gotska Sandön, I remembered the nights when Thorbjörn Stärner and I used to argue about what is determined and what you can decide of your own free will. It's only now that I recognize he was right in saying that character is pretty well fixed by the time you reach your teens. Eisen sought happiness – or meaning, if you like – by repeating one and the same project over and over again but in different forms.

Portraits of Washington was by any standards an undertaking on the grand scale, during which the author corresponded with museums and private collectors all over the world. It is difficult to get an idea of how many portraits he actually found, but it must have run into the thousands. He managed to track down a dozen portraits by the Swedish painter Adolph Ulric Werthmüller alone, all of which he then described in detail, along with their genesis, ownership and history. I remained sitting there for days, amazed and captivated, until finally liberated by a commentary on page 1,021:

To be an expert simply means that one has experience. But there are different kinds of experience and quantity does not always have the same significance as quality. There are many experts and many people have great experience, but no one is infallible. As far as the general public is concerned the person who rejects and ridicules most often is the one regarded as having most knowledge. How else would he be able to see that others are wrong?

It is a common belief that approval does not demand knowledge since anyone and everyone can express approval without knowing anything about the subject. But to reject something is a different matter and demonstrates knowledge. This author is of the opinion that there aren't any experts when it comes to art. Our present knowledge may be worth paying attention to, but it will be questioned within a couple of years and before a generation has passed it will be regarded as no more than 'the delusions of a past age'. In cases where there is doubt and disagreement the best way is to seek for the provenance and to assess it in a rational and unprejudiced way. All knowledge is at best approximate.

I recognized that tone. The topic itself may be an echo of René Malaise – that idiot – but the tone is reminiscent of Charles Darwin in the autumn of his years. Let other people deal with the arguments; the important thing is to do the best one can. And if you can enjoy yourself while doing so, then that is a plus.

I never really got to know Gustaf Eisen, but there were times when I thought I knew what he was up to and why. He wrote to Stuxberg as early as January 1875, telling him his plan.

> *Unless I'm struck by really bad luck I have come to a decision, which is as follows: 1. Never to seek a permanent post with a salary. 2. To make myself financially independent. 3. To study for the sake of my own greater enjoyment and for the advancement of science, not to promote the interests of the few. 4. To enjoy as much of life as possible in an open-handed sort of way, by which I mean doing what I please, travelling where I want and studying what I will.*

He was twenty-seven years old at that point and, for the most part, he managed to stick to his guns. At least he bore his failures with dignity. He wasn't really able to achieve economic independence and was forced to work for others quite a bit, but somehow or other he usually succeeded in getting his funders to dance to his tune.

In an interview somewhere or other he said that he had few regrets about his life. That's what we all say, of course. But when he was drawing up his final accounts, there was one thing he regretted and that was his exile: he regretted having left Sweden and become an American. Not to the extent of becoming embittered – he had no time for that sort of thing – but I believe that the loneliness of old age made him long for home.

At the end of the 1930s he wrote a long letter to Nils Dahlbeck, the then secretary of the Swedish Association for the Protection of Nature, about an ancient spruce tree that he remembered from the summer of 1854. It stood close to the manse in Harg,

Uppland, and he wanted to know whether it was still there. He described the place very precisely and attached a sketch of the tree as he saw it in his memory. Dahlbeck answered respectfully that he had unfortunately not been able to locate this special tree: 'It seems likely that it has gone. Nature in Sweden is now being remodelled at an unbelievable speed. What can be saved by those of us who are attempting to preserve the riches of nature are no more than fragments.'

Eisen had saved the greatest trees in the world, yet it was the spruce trees of his childhood that were in his mind's eye when he was a very old man living on Park Avenue. I think I can understand why. You can assimilate and love many things even in far-off countries, but the most profound feelings for forests and for meadows, for birds and for the drifting scent of dog roses at twilight in summer, are nevertheless the things that remain with you when old age, itself something of a foreign country, comes. In a different interview Eisen said, with unusual heat for him: 'I know that I'm old and it makes me really angry.'

. . .

It is typical that the grand apartment in Manhattan he occupied for the last two decades of his life was not his own. Fahim Kouchakji and his young American wife Evelyn lived there, and they looked after Eisen as if he were an aged relative. When he was in his nineties he was knocked down by a truck during one of his morning walks and his hip was crushed so badly that no one thought he would walk again, but he did, of course, after a couple of months in plaster.

He couldn't run around like a young lad any more, but he was in amazingly good spirits when Erik Wästberg visited to interview him for *Vecko-Journalen* shortly afterwards. Wästberg wrote: 'He is a Renaissance man, a man of universal learning, a travelling researcher of imposing stature who belongs in past ages; but even a modern journalist with ephemeral and negativistic tendencies cannot approach his desk without respect – a desk on which a Syrian temple cat is tiptoeing around among the manuscripts.'

So we can say that Gustaf Eisen ended his days looking after a cat on Park Avenue. Kouchakji and his wife also had a parrot that he looked after when they were away on their frequent travels all over the world. And the squirrels in the park shouldn't be forgotten; they seem to have been his very best friends.

In all the years he lived in New York he spent three hours every morning – from six to nine – in Central Park, where he walked, thought and, above all, fed the squirrels and the birds. One of the most interesting manuscripts I found in Uppsala was an impassioned 25-page appeal to the mayor of New York about the slummification of Central Park. I don't know whether it was published or put to some other use, but we stopped off in the city on our way home from California and took a room on the Upper West Side. We were on our way to the Metropolitan Museum to look at the Antioch Chalice, and since we were in the area I took the opportunity to stroll around the park for a couple of days.

Central Park is an island in the city, a true paradise. If I was permanently resident in Manhattan, I, too, would certainly go there every morning. As I said, I don't know what part Eisen played in improving the park, but I was able to tick off point by point that what he wanted had been done, even if it had perhaps been done posthumously. The story of the nest-boxes does, however, suggest that he had some influence on developments while he was still alive.

It was during the last years of his life.

His days and his evenings were fully occupied with learning to read cuneiform writing, a skill that came in handy for the book on Babylonian, Mesopotamian and other cylinder seals he was working on at the time and that was published in 1940 by the University of Chicago Oriental Institute. His mornings, however, were spent as usual in Central Park. He knew over a hundred squirrels as individuals – by name – and we can assume that the friendships were mutual. He must have been one of the real characters of the park and clearly highly respected as such, since we are told that the park administration hired a limousine every spring to drive Eisen around the park so that he could point out the trees in which they should fix new nest-boxes. What more can a man ask for?

Chapter 18

My Flies Leave the Island

My own search for a clearly circumscribed field that would provide a purpose in life eventually led me to a slim volume, *The Fauna of Lund Cathedral*. A scientist, admirable for the propriety of his objectivity, reports what he found in May 1936 while making an inventory of the sacred edifice. Everything from bats in the belfry and woodlice in the sacristy down to the springtails of various kinds he found in the mould and litter in the darkest corners of the crypt. He had even fished for plankton in the thousand-year-old well.

'Perfect!' I thought. Herein, perhaps, lay the mysterious and seemingly impossible gathering together of everything that I had been seeking by candlelight and lantern-light for so many years. Full of curiosity and without a further thought, I sprinted through the gateway of his narrative, net at the ready, rather like a twelve-year-old in the summer dusk when ghost moths are at last on the wing over the nettles and meadowsweet in an overgrown market garden between sea and forest.

Now! Now!

And then something unexpected happened. Something I'd been longing for, too, I noticed.

I heard the gate closing behind me. I stood there alone in the darkness, listening. And the only thing I sensed, however long and hard I concentrated, was the smell of dust. I really tried. Everything fitted and this path was so obviously the right one. The researcher in question had even contributed to science by discovering a hitherto unknown springtail in the cellars of the cathedral, a species that, because of where it was found, was later christened *Pseudosinella religiosa*. Not even that made my heart beat faster. The desire had left me.

I turned on my heel and walked back out.

It was my last attempt, cut short before it had even started.

Once out in the fresh air it struck me that vanity and ambition, my old urge to be best, had begun to abate somewhat. The fear of being forgotten and excluded was also diminishing. It wasn't so much that my interest was beginning to focus solely on the romance of it all, though there is no doubt that its beauty was becoming more and more interesting to me. I don't know what to call it. All these species were a language, and now I knew so much vocabulary by heart that I could pay more attention to the grand narrative and move beyond the island of security provided by defined limits.

Something had come to an end; it was just that I hadn't noticed.

It was only when my flies left the island that I realized it was all over. All deference to natural history, but what is it in comparison to art?

. . .

I had always assumed that this story would have to be told quickly.

That's not how it turned out.

In fact, nothing turned out as intended.

Unforeseen events occurred one after the other in quick succession, and in the end my collection of flies – my hoverflies from the island in the archipelago – was transformed before my very eyes into something quite different. It wasn't anything I'd planned, not consciously anyway. Chance circumstances, nothing else … perhaps?

All of a sudden the flies were no longer mine. The collection, ten years of my life and several thousand dried flies pinned in serried ranks like a military parade on Red Square, began to live a life of its own. Now the stage was theirs, the spotlight was on them. They would become art, contemporary art. I find it hard to imagine an idea more sick. Fortunately it wasn't me who came up with it.

There was something about the very nature of collecting that attracted artistic people: human behaviour and passions in general, concentration and precision. I assume this was why there came a day when, as an expert, I was invited to participate in a seminar at the Modern Museum in Stockholm, the theme of the event being the psychology of collecting. The organizers' idea was that a psychoanalyst, an art expert and I would discuss the topic in front of an audience. It sounded dangerous, but I couldn't say no: my vanity hadn't faded to that extent.

I was very suspicious of the psychoanalyst, not so much from conviction as from loyalty to Vladimir Nabokov, whose books I was crazy about at the time. He disliked both Freudians and Symbolists, as I did, too. You have to stick together! He objected to attempts to find some kind of Symbolism in the fact that he collected butterflies, and if anyone tried to portray his enthusiasm as an expression of other – darker – impulses he became both angry and sad.

During the Second World War, Nabokov worked as a professional entomologist, a specialist in blues, at the Harvard Museum of Comparative Zoology in Boston, but it had nothing at all to do with his novels. Not directly anyway. There was a distant, rather more indirect connection, as he explained on one occasion, between what he called 'the precision of poetry and the excitement of pure science', which no one really understood at the time – and he probably didn't intend them to.

Anyway, he convinced me to view Freudians as altogether too bound by authority and consequently rather ridiculous, so once I was up there on the platform at the Modern Museum I went at it hard and set traps for my interlocutor. My strategy soon fell apart because the psychoanalyst in question was a charming woman who made a lot of thought-provoking points on the topic of the importance of childhood. We left the arts fellow out of it and he was sufficiently unworldly as not to disturb us when we started discussing how many shoes you have to own in order to qualify as a shoe fetishist. We ended up thoroughly enjoying ourselves.

And that is more or less when my career as an artist began. Or, more accurately, the fly collection's career as art.

It happened as follows. A couple of years after this seminar it was decided that the Nordic Pavilion at the 2009 Venice Biennale would be curated by two artists resident in Berlin, one a Dane and the other a Norwegian. They were two artists with worldwide reputations among experts in the field, and they sat respectably enough at about 150th in the world rankings for contemporary artists – anywhere in the first thousand is regarded as sensationally good. As for me, I'd never heard of them. They decided early on that the exhibition would be called *The Collectors* and would consist of works of art that in one way or another reflected the joys and the madness of collecting. So now they were looking for suitable subjects.

One of the curators at the Modern Museum was helping them with their search and happened to recall that the museum had been visited a couple of years earlier by an expert on collecting, a man who had talked about flies – shoes, too, possibly – at a joint seminar with a Freudian. This fly-expert might have something to contribute. Contact was made and we met in the museum one autumn evening, talked about the exhibition, exchanged pleasantries and tentatively probed each other's expertise. Strange as it may seem, the result of this meeting was a request to exhibit my flies in Venice. That, as I write, is where the flies are at the moment, the whole collection! And that in my eyes is the ultimate and most definitive proof that international contemporary art is finished and bankrupt.

I did not, however, have a moment's hesitation. My collection of flies, stored in 144 plastic boxes, which, in turn, are arranged in nine large aluminium cases with glass lids, is really worth seeing. It may not be art, but, unlike all my other works, and I mean all, right from the very first newspaper article I wrote for *Västerviks-Tidningen* when I was seventeen, unlike all the books and the rest, my only motive for making the fly-collection was that it gave me pleasure. (That first article, by the way, was about a centuries-old oak, hollow as a drum and close to Gränsö Canal, which the uncultured barbarians of the electricity company had chopped down.)

Seen in that light, the fly-collection is a distillation of unthinking happiness. If it tells us anything at all it is perhaps that freedom starts when we take a step to one side and, if only for a moment, do something that has no purpose beyond itself, something that is not done in vain pursuit of respect, appreciation, power, money, love, fame or honour.

My *Callicera* went with the collection, too, even though it's the only genus of flies I collect nowadays. It didn't seem right to keep them at home when all the others were taking a trip to Venice. All or nothing.

As I was trundling my collection down to the quay in a wheelbarrow on an overcast day in March along a road still covered in greyish-brown slush, I was granted an unexpected insight into the form language of the Symbolists. The cases were on their way to the mainland, to my friend the carpenter, a man who had once earned a modicum of recognition by making a complete list of the insect life on a solitary windblown stump in the middle of a clear felling. Now he was going to link these nine cases in threes in order to make a triptych several metres long that could be hung on the wall at the Biennale.

I saw myself from above, so to speak, on the road, in the slush.

The cases were heavy and the boat was waiting.

They travelled south with an insurance value greater than the price we paid when we bought the house on the island. And perhaps I'll never see them again.

I didn't attend the opening. I can't be doing with too much attention, nor with champagne for that matter. But my representative who was present at the occasion has given me an account of how Queen Sonja of Norway, surrounded by the world's press, opened the exhibition while all my flies did their very best to stand to attention and look a little superior. They are to hang there for six months, and when the exhibition closes in November there won't be any flies in the whole wide world that have been so close to so many beautiful women.

Men, too, of course. The curators of the show don't make any secret of the fact that this year's Nordic Pavilion is an essentially male affair, positively homoerotic. Near my hoverflies, for instance, there hangs a whole series of works by the obscene artist Touko Laaksonen (1920–91), alias 'Tom of Finland'. Throughout his whole career he insisted on depicting sailors and American police constables with muscular forearms and equally pumped-up sexual organs. Contemporary art aims to be challenging. Provocative. Or that's what it is supposed to be, but let's hope it will soon be a thing of the past.

It occurs to me that back in the 1980s there was a punk band in Gothenburg that went by the name of Tattooed Copcocks. Their music was pretty much as you might expect. The only good thing about them was their name, which really was a bit of a gem for anyone with the slightest interest in the creative possibilities of language. Presumably it was meant to be provocative, but it didn't work because provocation as an artistic method had by then become institutionalized, had become something that the academy insisted on.

What has happened is that the angry cutting edge of the visual arts with its obsessive breaking of taboos has become a comfort zone for artists who neither can nor dare follow their own road. It's not hard to be thought-provoking and even simple souls can manage it, even people in advertising. Beauty, on the other hand, is a different matter: for an artist whose ambitions extend further than the local arts centre to try to approach beauty these days demands the kind of courage that is utterly alien to all those challenging, ironically cynical, ground-breaking provocateurs.

When flies are exhibited on the wall in Venice, the end is nigh, very nigh.

I wouldn't be surprised if some expert took them to be dots, just dots, like a raster image, maybe even some kind of innovative pointillism. If by any chance a prospective buyer, his fortune made in a manner I'd rather remain ignorant of, comes up with an offer I shall sell at the highest possible price. And if anyone can be bothered to listen I'll give an account of the unit cost one by one. Not that it's likely to be much more than a thousand or so per fly, though for certain specimens that would be bargain-basement pricing: acquiring a *Callicera aurata* is not something that's granted to everyone.

But it's a case of all or nothing.

No one would be happier than me if my flies were to serve as final proof that contemporary visual arts have become a market-stall, a circus in which the elephants were long ago replaced with a growing multitude of clowns.

Died-in-the-wool atheists who have nevertheless retained a profound sense of the religious are frequently pressed to explain what it is that offers them spiritual experiences comparable to the incense, myths and uplifting tales of the old religions. In response they usually refer to certain poets or to classical music, even to quite modern music sometimes, but seldom, very seldom, do they mention the contemporary visual arts. If we think of them as an analyst's couch, the visual arts may not be all bad, but that is the only way. I know I'm being unjust and that there are exceptions, but the visual arts have a long way to go, not backwards but in any other direction at all, in order to rediscover the beauty that literature, music, dance and architecture have never abandoned.

We can, however, say one thing in praise of the curators of *The Collectors*, which is that it is an honest salon, at least in the sense that it makes no attempt to mince words when it comes to the significance of love, even if it that love is only in the crassly biological form of reproductive reflexes. Crude, perhaps, but honest. There is something to be said for making one's intentions sufficiently clear that they can be seen by all who want to see them, whether the intentions are of an erotic kind or simply a striving to navigate around the deep pit of loneliness.

. . .

I now possess a bat detector, a pocket-sized piece of equipment that electronically converts the ultrasound of bats to fully audible frequencies. Every lonely individual ought to have such a device, as indeed should people too shy to approach their neighbours and invite them to take a stroll in the dark.

It's not always easy to do that, is it?

There you are, sitting on someone's veranda on a summer's evening when the conversation begins drifting in the direction of cinema – a sure indication that the party is running out of steam and will soon peter out in disappointment. You switch on your detector. There are bats more or less everywhere and you soon learn to distinguish the sounds of the common species. You don't have to make a great fuss about it, just say, as if in passing, as if talking to yourself:

'Sounds like a lot of soprano pipistrelles are flying tonight.'

Or whatever it might be that you can hear.

The keenest analysis of Fellini isn't going to survive a gambit like that. Ingmar Bergman himself might as well give up and go home to bed. You now have an open field before you and the night itself is smiling. Really! And a little later, when you say you want to stretch your legs and take a stroll in the moonlight, perhaps as far as the shore to get a bearing on the sounds of the Daubenton's bats, you'll find you won't be going alone.